An Introduction to Linear Regression and Correlation

A series of books in psychology

Editors:
Richard C. Atkinson
Gardner Lindzey
Richard F. Thompson

An Introduction to Linear Regression and Correlation

Second Edition

Allen L. Edwards

University of Washington

W. H. Freeman and Company
New York

Library of Congress Cataloging in Publication Data

Edwards, Allen Louis.
    An introduction to linear regression and correlation.

    (A Series of books in psychology)
    Includes index.
    1. Regression analysis.   2. Correlation (Statistics)
I. Title.   II. Series.
QA278.2.E3   1984   519.5′36   83-20777
ISBN 0-7167-1593-7
ISBN 0-7167-1594-5 (pbk.)

Printed in the United States of America

10 9 8 7 6 5 4 3 2

For David

Contents

Preface

This book was written for students of one of the behavioral sciences, psychology, but students of the other behavioral sciences may also find it of interest. What I have tried to do in the book is to provide the student with a more detailed and systematic treatment of linear regression and correlation than that ordinarily given in either a first or a second course in applied statistics for students of psychology.

The book has been written at a level that can be understood by any student with a working knowledge of elementary algebra. I have *not* assumed that the reader has already been exposed to a first course in statistics. Many of the topics traditionally covered in the first course are not essential to an understanding of linear regression and correlation and those that are essential have, I believe, been briefly but adequately covered in this book. Consequently, the book may be used as a text in either a first or a second course in statistics. If students are exposed to only a single course in statistics, one in which a more traditional book is used as a text, then this book might be considered as supplementary reading to provide a more detailed coverage of the topics of linear regression and correlation.

Since the publication of the first edition of the book in 1976, I have taught each year a two credit–noncredit course in which the book has been used as a text. The only prerequisite for the course is that the student have a working knowledge of algebra. Most of my students have also been exposed, at some time in the past, to a course in elementary statistics. The students come from a wide variety of academic disciplines, but primarily from the behavioral sciences. Many are psychology majors. Others come from education, sociology, educational psychology, business administration, nursing, communications research, and speech. I have also had students from musicology.

Class time is spent primarily in clarifying any aspect of the text that students have difficulty in understanding, including any proofs that they may have difficulty in following. When time permits, I discuss various other topics, not covered in the text, that are relevant to linear regression. I also provide students with copies of computer output for one or more of the examples discussed in the text and go over the additional information provided by this output. For this purpose, I have primarily used Minitab and SPSS programs.

In order to receive credit, students must complete all the exercises at the end of each chapter. In addition, I often give take-home quizzes which are not graded but which provide a basis for class discussion. I encourage students, who so desire, to use Minitab or SPSS with the problems in the exercises. I do insist, however, that they make additional calculations with a hand calculator. For example, computer output does not show how a regression sum of squares, regression coefficients, or standard errors of regression coefficients were obtained. In all the exercises needing it, I have provided $(X'X)^{-1}$. So, even though students may use the computer, they must still do calculations that involve $(X'X)^{-1}$ to find regression coefficients, standard errors, and so on.

My class is informal and nonthreatening, possibly because it is offered as a credit–noncredit course. As a result, I have found that students are not hesitant to ask questions or to respond to questions. We have a very interactive student-instructor relationship. The changes that have been made in this edition reflect, in great part, the results of this interactive relationship and the interests of my students. They have wanted more about multiple regression and this revision provides that. Four chapters in the first edition have been eliminated and have been replaced by four new chapters. The new chapters give considerable more attention to multiple regression than did the single chapter on this topic in the first edition.

The book begins at a very elementary level with the equation of a straight line. There is a continuity of development in each of the successive chapters. The second chapter is concerned with values of a dependent variable Y that are subject to random variation. The student is shown how the method of least squares can be used to find a line of best fit, and the residual variance and standard error of estimate as measures of the variation of the Y values about the line of best fit are introduced.

Chapter 3 deals with the correlation coefficient as a measure of the degree to which two variables are linearly related. The relationship of the correlation coefficient to the residual variance and standard error of estimate is explained. The coefficients of determination and nondetermination are discussed. Various factors that may be related to the magnitude of the correlation coefficient are discussed in Chapter 4. In Chapter 5 the phi coefficient, point biserial coefficient, and the rank order coefficient are shown to be merely special cases of the correlation coefficient.

Chapter 6 begins with a discussion of a model for a correlational problem. There is a brief discussion of tests of significance and of the four major distributions—the normal, t, F, and χ^2 distributions—used in making such tests. The treatment is nonmathematical and intuitive, and it is at a level that can be understood by the beginning student. The t test of the null hypothesis that the population correlation is zero and Fisher's z_r transformation for the correlation coefficient are then discussed. The standard normal distribution test of the difference between two independent correlation coefficients is illustrated, along with the χ^2 test for the homogeneity of several independent values of the correlation coefficient. Chapter 7 is concerned with tests of significance for the special cases of the correlation coefficient; Chapter 8 deals with tests of significance for regression coefficients.

Chapter 9 provides an example of a multiple regression problem with two X variables. Chapter 10 is concerned with matrices and a little bit of matrix algebra. In Chapter 11 the general case of multiple regression is discussed using as an example a problem with three X variables. Chapter 12 analyzes the same example in terms of standardized variables. Multiple regression with dummy and effect coding is discussed in Chapter 13. The book concludes with a discussion of coefficients for orthogonal polynomials.

I consider the exercises at the end of each chapter an integral part of the book. They are designed to test the reader's understanding of the material covered in that chapter. In all cases where it is needed, I have provided the reader with $(X'X)^{-1}$. Additional calculations involving $(X'X)^{-1}$ can easily be accomplished using a hand calculator.

In some of the exercises I have asked for a proof. In some cases the proof has already been given in the text. In other cases the proof is provided in the Answers to the Exercises. In still other cases the proof is left to the student.

Tables II and IV in the Appendix have been reprinted from R. A. Fisher, *Statistical Methods for Research Workers* (14th ed.), Copyright 1972 by Hafner Press, by permission of the publisher. Table III is reprinted from Enrico T. Federighi, Extended tables of the percentage points of Student's *t* distribution, *Journal of the American Statistical Association,* 1959, *54,* 683–688, by permission of the American Statistical Association. Table VI has been reprinted from George W. Snedecor and William G. Cochran, *Statistical Methods* (6th ed.), Copyright 1967 by Iowa State University Press, Ames, Iowa, by permission of the publisher.

Allen L. Edwards
Seattle, Washington

An Introduction to Linear Regression and Correlation

One

The Equation
of a Straight Line

1.1 Introduction

Many experiments are concerned with the relationship between an independent variable X and a dependent variable Y. The values of the independent variable may represent measures of time, number of trials, varying levels of illumination, varying amounts of practice, varying dosages of a drug, different intensities of shock, different levels of reinforcement, or any other quantitative variable of experimental interest. Ordinarily, the values of the X variable *in an experiment* are selected by the experimenter and are limited in number. They are usually measured precisely and can be assumed to be without error. In general, we shall refer to the values of the X variable in an experiment as fixed in that any conclusions based on the outcome of the experiment will be limited to the particular X values investigated.

For each of the X values, one or more observations of a relevant dependent Y variable are obtained. The objective of the experiment is to determine whether the Y values (or the average Y values, if more than one observation is obtained for each value of X) are related to the X values. In this chapter we shall be concerned with the case where the Y values are linearly related to the X values. By "linearly related" we mean that if the Y values are plotted against the X values, the resulting trend of the plotted points can be represented by a straight line. If the Y values are linearly related to the X values, then we also want to determine the equation for the straight line. We may regard this equation as a rule that relates the Y values to the X values.

1.2 The Equation of a Straight Line

Consider the values of X and Y shown in Table 1.1. What is the rule that relates the values of Y to the values of X? Examination of the pairs of values will show that for each value of X, the corresponding value of Y is equal to $-.4X$. We may express this rule in the following way:

$$Y = bX \tag{1.1}$$

where $b = -.4$ is a constant that multiplies each value of X. If each value of Y in Table 1.1 were exactly equal to the corresponding value of X, then the value of b would have to be equal to 1.00. If each value of Y were numerically equal to X, but opposite in sign, then the value of b would have to be equal to -1.00.

Now examine the values of X and Y in Table 1.2. The rule or equation relating the Y values to the X values in this case has the general form

$$Y = a + bX \tag{1.2}$$

where b is again a constant that multiplies each value of X and a is a constant that is added to each of the products. For the values of X and Y given in Table

TABLE 1.1 $Y = -.4X$		TABLE 1.2 $Y = 2 + .3X$		TABLE 1.3 $Y = 3 + .5X$	
X	Y	X	Y	X	Y
10	−4.0	10	5.0	10	8.0
9	−3.6	9	4.7	9	7.5
8	−3.2	8	4.4	8	7.0
7	−2.8	7	4.1	7	6.5
6	−2.4	6	3.8	6	6.0
5	−2.0	5	3.5	5	5.5
4	−1.6	4	3.2	4	5.0
3	−1.2	3	2.9	3	4.5
2	− .8	2	2.6	2	4.0
1	− .4	1	2.3	1	3.5

1.2, the value of b is equal to .3 and the value of a is equal to 2. Thus when $X =$ 10, we have $Y = 2 + (.3)(10) = 5$. When $X = 8$, we have $Y = 2 + (.3)(8) = 4.4$.

Both (1.1) and (1.2) are equations for a straight line. For example, we could take any arbitrary constants for a and b. Then for any given set of X values we could substitute in (1.2) and obtain a set of Y values. If these values of Y are plotted against the corresponding X values, the set of plotted points will fall on a straight line.

1.3 Graph of $Y = a + bX$

Table 1.3 gives another set of X and Y values. Let us plot the Y values against the corresponding X values. The resulting graph will provide some additional insight into the nature of the constant b that multiplies each value of X as well as the nature of the constant a that is added to the product. In making the graph we set up two axes at right angles to each other. It is customary to let the horizontal axis represent the independent or X variable and the vertical axis represent the dependent or Y variable. We need not begin our scale on the X and Y axes at zero. We may begin with any convenient values that permit us to plot the smallest values of X and Y. In Figure 1.1, for example, we begin the X scale with 0 and the Y scale with 2.0. Nor is it necessary that the X and Y scales be expressed in the same units, as they are in Figure 1.1.

You will recall that a pair of (X, Y) values represents the coordinates of a point. To find the point on the graph corresponding to (10, 8.0), we go out the X axis to 10 and imagine a line perpendicular to the X axis erected at this point. We now go up the Y axis to 8.0 and imagine another line perpendicular to the Y axis erected at this point. The intersection of the two perpendiculars

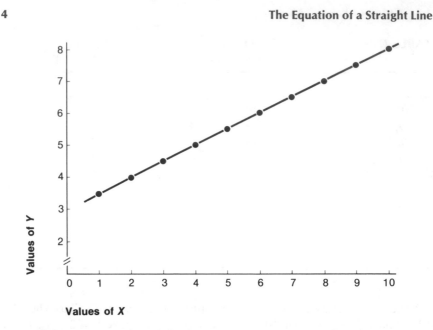

FIGURE 1.1   Plot of the (X, Y) values given in Table 1.3.

will be the point (10, 8.0) on the graph. It is obviously not necessary to draw the perpendiculars in order to plot a set of points.

1.4 The Slope and Intercept of a Straight Line

It is clear that the points plotted in Figure 1.1 fall along a straight line. The equation of this line, as given by (1.2), is

$$Y = a + bX$$

What is the nature of the multiplying constant b? Note, for example, that as we move from 7 to 8 on the X scale, the corresponding increase on the Y scale is from 6.5 to 7.0. An increase of one unit in X, in other words, results in .5 of a unit increase in Y. The constant b is simply the rate at which Y changes with unit change in X.

The value of b can be determined directly from Figure 1.1. For example, if we take any two points on the line with coordinates (X_1, Y_1) and (X_2, Y_2), then

$$b = \frac{Y_2 - Y_1}{X_2 - X_1} \tag{1.3}$$

Substituting in (1.3) the coordinates (2, 4.0) and (3, 4.5), we have

$$b = \frac{4.5 - 4.0}{3 - 2} = .5$$

In geometry (1.3) is known as a particular form of the equation of a straight line, and the value of b is called the slope of the straight line.

The nature of the additive constant a in (1.2) can readily be determined by setting X equal to zero. The value of a must then be the value of Y when X is equal to zero. If the straight line in Figure 1.1 were to be extended downward, we would see that the line would intersect the Y axis at the point $(0,a)$. The number a is called the Y-intercept of the line. In our example, it is easy to see that the value of a is equal to 3. If a straight line passed through the point $(0,0)$, then a would be equal to zero, and the equation of the straight line would be $Y = bX$.

1.5 Positive and Negative Relationships

We may conclude that if the relationship between two variables is linear, then the values of a and b can be determined by plotting the values and finding the Y-intercept and the slope of the line, respectively. A single equation may then be written that will express the nature of the relationship. When the value of b is positive, the relationship is also described as positive; that is, an increase in X is accompanied by an increase in Y and a decrease in X is accompanied by a decrease in Y. When the value of b is negative, the relationship is also described as negative. A negative relationship means that an increase in X is accompanied by a decrease in Y, and a decrease in X is accompanied by an increase in Y. When two variables are positively related, the line representing the relationship will extend from the lower left of the graph to the upper right, and the slope of the line will be positive. When the relationship is negative, the line will extend from the upper left of the graph to the lower right, and the slope of the line will be negative.

Exercises

1.1 Find the values of a and b in the equation $Y = a + bX$ for the following paired (X,Y) values:

X	Y
1	2.2
2	2.6
3	3.0
4	3.4
5	3.8

1.2   Find the values of a and b in the equation $Y = a + bX$ for the following paired (X,Y) values:

X	Y
1	−5.4
2	−5.8
3	−6.2
4	−6.6
5	−7.0

1.3   Find the values of a and b in the equation $Y = a + bX$ for the following paired (X,Y) values:

X	Y
2	10.6
4	11.2
6	11.8
8	12.4
10	13.0

1.4   Find the values of a and b in the equation $Y = a + bX$ for the following paired (X,Y) values:

X	Y
1.0	4.8
1.5	4.7
2.0	4.6
2.5	4.5
3.0	4.4

1.5   Find the values of a and b in the equation $Y = a + bX$ for the following paired (X,Y) values:

X	Y
20	0
16	2
10	5
6	7
0	10

1.6 Find the values of Y in the equation $Y = 2 + 3X$ for the following values of X: -3, -2, 0, 1, and 5.

1.7 Find the values of Y in the equation $Y = -3 + 1.5X$ for the following values of X: -3, -2, 0, 1, and 5.

1.8 Find the values of Y in the equation $Y = -1.5 + 2X$ for the following values of X: -3, -2, 0, 1, and 5.

1.9 Find the values of Y in the equation $Y = -1.5 - .5X$ for the following values of X: -3, -1, 0, 1, and 5.

1.10 Find the values of Y in the equation $Y = .5 - 2X$ for the following values of X: -3, -1, 0, 1, and 5.

1.11 Briefly explain the meaning of each of the following terms or concepts:

dependent variable
independent variable
slope of a straight line
Y-intercept
positive relationship
negative relationship

Two

The Regression Line of Y on X

2.1 Introduction

When a set of plotted points corresponding to the values of a Y variable and an X variable all fall precisely on a straight line, with nonzero slope, so that no point deviates from the line, the relationship between the two variables is said to be *perfect*. This means that every observed value of Y will be given exactly by $Y = a + bX$. Although, as we have pointed out in the previous chapter, the values of the independent variable X selected by an experimenter may be assumed to be measured precisely and are fixed, this will ordinarily not be true of the observed values of the dependent variable Y. When the Y values obtained in an experiment are plotted against the corresponding X values, the trend of the plotted points may be linear, but the plotted points will not necessarily fall precisely on any line that we might draw to represent the trend.

When the values of X are fixed and when the values of Y are subject to random variation, the problem is to find a line of best fit that relates Y to X. This line is called the *regression line* of Y on X and the equation of the line is called a *regression equation*. Because the values of Y will not necessarily fall on the regression line, we make a slight change in notation and write the regression equation as

$$Y' = a + bX$$

The actual or observed value of Y will then be

$$Y = a + bX + e$$

and

$$Y - Y' = Y - (a + bX) = e$$

will be an error or residual when Y is not equal to Y'.

2.2 The Mean and Variance of a Variable

Table 2.1 shows a set of 10 paired (X,Y) values, and Figure 2.1 shows the plot of the Y values against the X values. It is obvious that the values of Y tend to increase as X increases, but it is also obvious that the values of Y will not fall on any straight line that we might draw to represent the upward trend of the points.

The horizontal line $\overline{Y}$ in Figure 2.1 corresponds to the mean of the Y values. The *mean* value of a variable is defined as the sum of the observed values divided by the number of values and is ordinarily represented by a capital letter with a bar over it. For the mean of the Y values we have

$$\overline{Y} = \frac{\Sigma Y}{n} \tag{2.1}$$

TABLE 2.1   A Set of Ten Paired (X, Y) Values and the Values of $Y - \overline{Y}$, $(Y - \overline{Y})^2$, and $Y - Y'$

	(1) X	(2) Y	(3) $Y - \overline{Y}$	(4) $(Y - \overline{Y})^2$	(5) $Y - Y'$
	10	12	5.5	30.25	.01
	9	10	3.5	12.25	$-$ .77
	8	11	4.5	20.25	1.45
	7	9	2.5	6.25	.67
	6	7	.5	.25	$-$ .11
	5	5	-1.5	2.25	$-$ .89
	4	4	-2.5	6.25	$-$ .67
	3	2	-4.5	20.25	-1.45
	2	3	-3.5	12.25	.77
	1	2	-4.5	20.25	.99
Σ	55	65	0.0	130.50	0.00

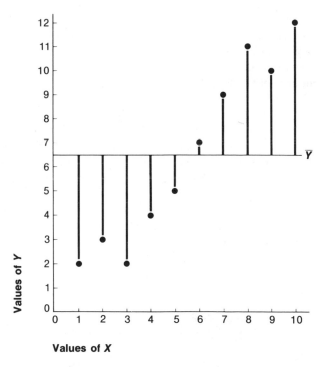

FIGURE 2.1   A plot of the (X, Y) values given in Table 2.1 showing the deviations of the Y values from the mean of the Y values.

where ΣY indicates that we are to sum the n values of Y. For the mean of the Y values in Table 2.1, we have

$$\overline{Y} = \frac{2 + 3 + 2 + 4 + \cdots + 12}{10} = 6.5$$

Similarly, for the mean of the X values, we have

$$\overline{X} = \frac{1 + 2 + 3 + 4 + \cdots + 10}{10} = 5.5$$

In Figure 2.1 vertical lines connect each plotted point and the mean (6.5) of the Y values. Each of these vertical lines represents a deviation of an observed value of Y from the mean of the Y values or the magnitude of $Y - \overline{Y}$. The values of these deviations are shown in column (3) of Table 2.1, and we note that the algebraic sum of the deviations is equal to zero, that is,

$$\Sigma(Y - \overline{Y}) = 0$$

This will always be true, regardless of the actual values of Y. For example, we will always have n values of $Y - \overline{Y}$ and, consequently, when we sum the n values, we obtain

$$\Sigma(Y - \overline{Y}) = \Sigma Y - n\overline{Y} = \Sigma Y - \Sigma Y = 0$$

because $\overline{Y}$ is a constant and will be subtracted n times. By definition, $\overline{Y} = \Sigma Y/n$ and, therefore, $n\overline{Y}$ will be equal to ΣY.

The squares of the deviations, $(Y - \overline{Y})^2$, are given in column (4) of Table 2.1, and for the sum of the squared deviations we have

$$\Sigma(Y - \overline{Y})^2 = 130.50$$

It can be proved,[1] for any variable Y, that $\Sigma(Y - \overline{Y})^2$ will always be smaller than $\Sigma(Y - c)^2$, where c is any constant such that c is not equal to $\overline{Y}$. For example, if we were to subtract the constant $c = 6.0$ from each value of Y in Table 2.1, then we would find that $\Sigma(Y - 6.0)^2$ is larger than $\Sigma(Y - 6.5)^2$ because $c = 6.0$ is not equal to $\overline{Y} = 6.5$.

If we divide the sum of squared deviations from the mean by $n - 1$, we obtain a measure known as the *variance*. The variance is ordinarily represented by s^2, and for the Y variable we have

$$s_Y^2 = \frac{\Sigma(Y - \overline{Y})^2}{n - 1} \tag{2.2}$$

---

[1] A simple proof requires a knowledge of the rules of differentiation as taught in the first course of the calculus. We want to find the value of the constant c that minimizes $\Sigma(Y - c)^2$. Expanding this expression we have

$$\Sigma Y^2 - 2c\Sigma Y + nc^2$$

Differentiating the latter expression with respect to c and setting the derivative equal to zero, we obtain $c = \overline{Y}$.

or, for the values shown in Table 2.1,

$$s_Y^2 = \frac{130.50}{10 - 1} = 14.50$$

The square root of the variance is called the *standard deviation*. Thus

$$s_Y = \sqrt{\frac{\Sigma(Y - \bar{Y})^2}{n - 1}}$$

and for the Y values given in Table 2.1 we have

$$s_Y = \sqrt{\frac{130.50}{10 - 1}} = 3.81 \tag{2.3}$$

    The variance and standard deviation are both measures of the variability of the Y values about the mean of the Y values. When the variance and standard deviation are small, the values of Y will tend to have small deviations from the mean; when the variance and standard deviation are large, the values of Y will tend to have large deviations from the mean.

2.3   Finding the Values of *a* and *b* in the Regression Equation

Figure 2.2 is another plot of the Y values in Table 2.1 against the X values. This figure shows the horizontal line through the mean Y value (6.5) and also the vertical line through the mean X value (5.5). These two lines will obviously intersect or cross at the point with coordinates (5.5, 6.5). This point is shown in Figure 2.2 as a small open circle and is labeled B. Now suppose we rotate the horizontal line represented by $\bar{Y}$ counterclockwise about the point B until we come to the dashed line Y'. This line, as drawn in Figure 2.2, is the regression line of Y on X. The equation of the regression line, as we shall shortly see, is

$$Y' = a + bX \tag{2.4}$$

where

$$a = \bar{Y} - b\bar{X} \tag{2.5}$$

and

$$b = \frac{\Sigma(X - \bar{X})(Y - \bar{Y})}{\Sigma(X - \bar{X})^2} \tag{2.6}$$

    Y', as given by (2.4), will no longer necessarily be equal to the observed value of Y corresponding to an observed value of X, if the relationship between Y and X is not perfect. We can, however, regard the value of Y' as a prediction

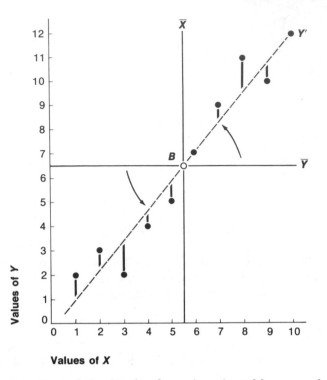

FIGURE 2.2   Plot of the (X,Y) values given in Table 2.1 and the line of best fit.

of the observed value of Y for the corresponding value of X. Then an error of prediction will be given by

$$Y - Y' = Y - (a + bX) = e \qquad (2.7)$$

These errors of prediction or deviations of Y from Y' are shown by the vertical lines in Figure 2.2, and the values of $Y - Y'$ are given in column (5) of Table 2.1. Note that $\Sigma(Y - Y') = 0$. We shall show why this is true later in this chapter.

The line Y' drawn in Figure 2.2 was determined by the *method of least squares*. This criterion demands that the values of a and b in equation (2.4) be determined in such a way that the *residual sum of squares*

$$\Sigma(Y - Y')^2 = \Sigma[Y - (a + bX)]^2 \qquad (2.8)$$

will be a minimum. It can be shown that the values of a and b in (2.4) that will make the residual sum of squares $\Sigma(Y - Y')^2$ a minimum must satisfy the following two equations[2]:

$$\Sigma Y = na + b\Sigma X \qquad (2.9)$$

$$\Sigma XY = a\Sigma X + b\Sigma X^2 \qquad (2.10)$$

If we divide both sides of (2.9) by *n* and solve for *a*, we have

$$a = \overline{Y} - b\overline{X}$$

where $\overline{Y} = \Sigma Y/n$ is the mean of the *Y* values and $\overline{X} = \Sigma X/n$ is the mean of the *X* values. If we now substitute $\overline{Y} - b\overline{X}$ for *a* in (2.10), we have

$$\Sigma XY = (\overline{Y} - b\overline{X})\Sigma X + b\Sigma X^2$$
$$= n\overline{XY} - bn\overline{X}^2 + b\Sigma X^2$$
$$= n\overline{XY} + b(\Sigma X^2 - n\overline{X}^2)$$

Solving for *b*, we have

$$b = \frac{\Sigma XY - n\overline{XY}}{\Sigma X^2 - n\,\overline{X}^2}$$

or

$$b = \frac{\Sigma XY - (\Sigma X)(\Sigma Y)/n}{\Sigma X^2 - (\Sigma X)^2/n} \qquad (2.11)$$

and *b*, as defined by (2.11), is called the *regression coefficient*.

The necessary values for calculating *b* are given in Table 2.2. Substituting these values in (2.11), we have

$$b = \frac{458 - (55)(65)/10}{385 - (55)^2/10} = \frac{100.5}{82.5} = 1.22$$

Because $\overline{X} = 55/10 = 5.5$ and $\overline{Y} = 65/10 = 6.5$, and we have just found that $b = 1.22$, we may substitute in (2.5) and find

$$a = 6.5 - (1.22)(5.5) = -.21$$

The regression equation for the data in Table 2.2 will then be

$$Y' = -.21 + 1.22X$$

Note now that if we predict a value of *Y* corresponding to the mean of the *X* values, we obtain

$$Y' = -.21 + (1.22)(5.5) = 6.5$$

---

[2]Again, a simple proof requires a knowledge of the rules of differentiation. By expanding the right side of (2.8), we obtain

$$\Sigma Y^2 - 2a\Sigma Y - 2b\Sigma XY + na^2 + 2ab\Sigma X + b^2\Sigma X^2$$

Differentiating this expression with respect to *a* and with respect to *b* and setting the two derivatives equal to zero, we obtain (2.9) and (2.10).

TABLE 2.2 **Finding a Line of Best Fit**

(1) X	(2) Y	(3) X²	(4) Y²	(5) XY	(6) Y'	(7) Y − Y'	(8) (Y − Y')²
10	12	100	144	120	11.99	.01	.0001
9	10	81	100	90	10.77	− .77	.5929
8	11	64	121	88	9.55	1.45	2.1025
7	9	49	81	63	8.33	.67	.4489
6	7	36	49	42	7.11	− .11	.0121
5	5	25	25	25	5.89	− .89	.7921
4	4	16	16	16	4.67	− .67	.4489
3	2	9	4	6	3.45	−1.45	2.1025
2	3	4	9	6	2.23	.77	.5929
1	2	1	1	2	1.01	.99	.9801
Σ 55	65	385	553	458	65.00	0.00	8.0730

which is equal to the mean of the Y values. The regression line will, therefore, pass through the point established by the means of the X and Y values or, in other words, the point with coordinates $(\overline{X}, \overline{Y})$. This will always be true for any linear regression line fitted by the method of least squares.

The predicted value of Y when X is equal to 3 will be

$$Y' = - .21 + (1.22)(3) = 3.45$$

and when X is equal to 9, the predicted value of Y will be

$$Y' = -.21 + (1.22)(9) = 10.77$$

The regression line will therefore pass through the points with coordinates $(5.5, 6.5)$, $(3, 3.45)$, and $(9, 10.77)$. If we draw a line through these points, this will be the regression line of Y on X, as shown in Figure 2.2. The predicted values of Y' for each of the other X values are given in column (6) of Table 2.2; these points all fall on the regression line shown in Figure 2.2.

2.4 The Covariance

The numerator of (2.11) is equal to the sum of the products of the deviations of the paired X and Y values from their respective means. Thus, we have

$$\Sigma(X - \overline{X})(Y - \overline{Y}) = \Sigma XY - \overline{Y}\Sigma X - \overline{X}\Sigma Y + n\overline{X}\overline{Y}$$
$$= \Sigma XY - n\overline{X}\overline{Y} - n\overline{X}\overline{Y} + n\overline{X}\overline{Y}$$
$$= \Sigma XY - n\overline{X}\overline{Y}$$

or

$$\Sigma(X - \overline{X})(Y - \overline{Y}) = \Sigma XY - \frac{(\Sigma X)(\Sigma Y)}{n}$$

It will be convenient to let $x = X - \overline{X}$ and $y = Y - \overline{Y}$. Then

$$\Sigma xy = \Sigma(X - \overline{X})(Y - \overline{Y})$$
$$= \Sigma XY - \frac{(\Sigma X)(\Sigma Y)}{n}$$

The sum of the products of the deviations of the paired X and Y values from their respective means, when divided by $n - 1$, is called the *covariance*, c_{XY}. Thus

$$c_{XY} = \frac{\Sigma xy}{n - 1} \tag{2.12}$$

and if c_{XY} is equal to zero, then b will also be equal to zero.

The denominator of (2.11) is equal to the sum of the squared deviations of the X values from the mean of the X values. Thus

$$\Sigma x^2 = \Sigma(X - \overline{X})^2$$
$$= \Sigma X^2 - 2\overline{X}\Sigma X + n\overline{X}^2$$
$$= \Sigma X^2 - \frac{(\Sigma X)^2}{n}$$

The sum of squared deviations divided by $n - 1$, as we pointed out earlier, is called the variance and is represented by s^2. Therefore, the variance of the X values is

$$s_X^2 = \frac{\Sigma x^2}{n - 1} = \frac{\Sigma(X - \overline{X})^2}{n - 1}$$

We thus see that the regression coefficient, as defined by (2.11), can also be written as

$$b = \frac{c_{XY}}{s_X^2} = \frac{\Sigma xy/(n - 1)}{\Sigma x^2/(n - 1)} = \frac{\Sigma(X - \overline{X})(Y - \overline{Y})}{\Sigma(X - \overline{X})^2}$$

2.5 The Residual and Regression Sums of Squares

Column (8) in Table 2.2 gives the values of $e^2 = (Y - Y')^2$. For the sum of these squared values we have $\Sigma e^2 = \Sigma(Y - Y')^2 = 8.0730$ and, in regression analysis, this sum of squares is called the *residual sum of squares*. Let us see if we can gain some additional insight into the nature of this sum of squares. By definition

$$Y' = a + bX$$

and because $a = \overline{Y} - b\overline{X}$, we also have

$$Y' = \overline{Y} - b\overline{X} + bX$$

Summing over the *n* values, we have

$$\Sigma Y' = n\overline{Y} - bn\overline{X} + b\Sigma X$$

Because $n\overline{X} = \Sigma X$, the last two terms on the right cancel, and we have

$$\Sigma Y' = \Sigma Y$$

The sum of the predicted values, $\Sigma Y'$, is thus equal to the sum of the observed values, ΣY, and the mean of the predicted values must therefore be equal to the mean of the observed values of *Y*. We see that this is true for the data in Table 2.2, where $\Sigma Y' = 65.0$ and $\Sigma Y = 65.0$. It also follows that the algebraic sum of the deviations of the observed values from the predicted values must be equal to zero. Thus

$$\Sigma e = \Sigma(Y - Y') = \Sigma Y - \Sigma Y' = 0$$

because we have just shown that $\Sigma Y' = \Sigma Y$.

We have just shown that a predicted value Y' can be written $Y' = \overline{Y} - b\overline{X} + bX$. Rearranging the last two terms, we have

$$Y' = \overline{Y} + bX - b\overline{X}$$
$$= \overline{Y} + b(X - \overline{X})$$

Then an error of prediction will also be given by

$$Y - Y' = (Y - \overline{Y}) - b(X - \overline{X})$$

Substituting $y = Y - \overline{Y}$ and $x = X - \overline{X}$ in the preceding expression, we have

$$Y - Y' = y - bx$$

Squaring and summing, we have, for the sum of the squared errors of prediction,

$$\Sigma(Y - Y')^2 = \Sigma y^2 - 2b\Sigma xy + b^2\Sigma x^2$$

But $b = \Sigma xy / \Sigma x^2$, and substituting in the preceding expression we have

$$\Sigma(Y - Y')^2 = \Sigma y^2 - 2\frac{(\Sigma xy)^2}{\Sigma x^2} + \frac{(\Sigma xy)^2}{\Sigma x^2}$$
$$= \Sigma y^2 - \frac{(\Sigma xy)^2}{\Sigma x^2} \tag{2.13}$$

Table 2.2 gives the necessary values for calculating

$$\Sigma y^2 = \Sigma(Y - \bar{Y})^2 = \Sigma Y^2 - \frac{(\Sigma Y)^2}{n}$$

Thus

$$\Sigma y^2 = 553 - \frac{(65)^2}{10} = 130.5$$

Then, because we have already found that $\Sigma xy = 100.5$ and that $\Sigma x^2 = 82.5$, we find that the residual sum of squares, as given by (2.13), is

$$\Sigma(Y - Y')^2 = 130.5 - \frac{(100.5)^2}{82.5} = 8.0727$$

which is equal, within rounding errors, to the sum of the squared errors of prediction shown in column (8) in Table 2.2.

Note now that if we subtract the residual sum of squares from $\Sigma(Y - \bar{Y})^2$, we have

$$\Sigma(Y - \bar{Y})^2 - \Sigma(Y - Y')^2 = 130.5000 - 8.0727 = 122.4273$$

The sum of squares we have just obtained is called the *regression sum of squares* and, as we now show, it is equal to $\Sigma(Y' - \bar{Y})^2$. We have

$$Y' = \bar{Y} + b(X - \bar{X})$$

Then

$$Y' - \bar{Y} = b(X - \bar{X})$$

and

$$\Sigma(Y' - \bar{Y})^2 = b^2 \Sigma x^2$$

but $b^2 = (\Sigma xy)^2/(\Sigma x^2)^2$ and substituting in the above expression, we have

$$\Sigma(Y' - \bar{Y})^2 = \frac{(\Sigma xy)^2}{\Sigma x^2} \qquad (2.14)$$

It will be convenient to refer to $\Sigma(Y - \bar{Y})^2$ as the *total sum of squares* or as SS_{tot}. Similarly, we can indicate the regression sum of squares as SS_{reg} and the residual sum of squares as SS_{res}. Then, as we have shown,

$$SS_{tot} = SS_{res} + SS_{reg} \qquad (2.15a)$$

or

$$\Sigma(Y - \bar{Y})^2 = \Sigma(Y - Y')^2 + \Sigma(Y' - \bar{Y})^2 \qquad (2.15b)$$

or

$$\Sigma y^2 = \left[\Sigma y^2 - \frac{(\Sigma xy)^2}{\Sigma x^2}\right] + \frac{(\Sigma xy)^2}{\Sigma x^2} \qquad (2.15c)$$

When the relationship between Y and X is perfect, either positive or negative, the residual sum of squares will be equal to zero and the regression sum of squares will be equal to the total sum of squares. If there is no linear relationship between Y and X, that is, if $b = 0$, then the regression sum of squares will be equal to zero and the residual sum of squares will be equal to the total sum of squares. In this instance, $Y' = \overline{Y} + b(X - \overline{X}) = \overline{Y}$ and the best prediction we could make for any observed value of Y would be $\overline{Y}$, regardless of the value of X associated with Y. It would be the best prediction in the sense that $\Sigma(Y - \overline{Y})^2$, the sum of squared errors of prediction, would be less than it would be for any other *single* value not equal to $\overline{Y}$.

By taking into account the linear relationship between Y and X, when one exists, we reduce the total variation in Y, that is, $\Sigma(Y - \overline{Y})^2$, by an amount equal to the regression sum of squares, $(\Sigma xy)^2/\Sigma x^2$. The residual sum of squares, $\Sigma y^2 - (\Sigma xy)^2/\Sigma x^2$, then measures the remaining variation in Y that cannot be accounted for by a linear relationship.

If you will look for a moment at Figure 2.2, which shows the regression line of Y on X, you will be able to see more clearly why $\Sigma(Y - Y')^2$ will be smaller than $\Sigma(Y - \overline{Y})^2$, if X and Y are linearly related. A horizontal line has been drawn through the mean of the Y values. The vertical deviation of each plotted point from this line represents the deviation of $Y - \overline{Y}$, and the sum of these squared deviations is equal to $\Sigma(Y - \overline{Y})^2 = 130.5$. If the horizontal line through the mean of the Y values is now rotated counterclockwise about the point B, where the coordinates are $(\overline{X}, \overline{Y})$, then the sum of squared deviations from the line will become smaller and smaller until the line coincides with the regression line—line Y' in Figure 2.2. The sum of squared deviations from the regression line is $\Sigma(Y - Y')^2$, and $\Sigma(Y - Y')^2$ will be smaller than $\Sigma(Y - \overline{Y})^2$, if there is any linear relationship between X and Y. In the present example, we have $\Sigma(Y - Y')^2 = 8.07$, whereas $\Sigma(Y - \overline{Y})^2 = 130.5$. We see that $\Sigma(Y - Y')^2$ represents a considerable reduction in the sum of squared errors of prediction relative to $\Sigma(Y - \overline{Y})^2$.

It is the second variable X that makes the regression line and $\Sigma(Y - Y')^2$ meaningful. As long as the Y measures are considered alone, the best predicted value of Y for any single value of X would be the horizontal line, or the mean of the Y values. But when there is regression of Y on X, we find that different values of Y' are associated with different values of X. These associated values become our predictions when we know the relationship between the two variables.

2.6 The Residual Variance and Standard Error of Estimate

If we divide the residual sum of squares by $n - 2$, we obtain a measure known as the *residual variance*. Thus

$$s_{Y.X}^2 = \frac{\Sigma(Y - Y')^2}{n - 2} = \frac{\Sigma y^2 - (\Sigma xy)^2/\Sigma x^2}{n - 2} \qquad (2.16)$$

and for the data in Table 2.2, we have

$$s_{Y.X}^2 = \frac{8.07}{10 - 2} = 1.01$$

The residual variance is, as we have pointed out, a measure of the variation of the Y values about the regression line. The dot separating the Y and X subscripts indicates that the regression line is that of Y on X; that is, that we are predicting Y values from the corresponding X values.

The square root of the residual variance is called the *standard error of estimate*. Thus

$$s_{Y.X} = \sqrt{\frac{\Sigma(Y - Y')^2}{n - 2}} \tag{2.17}$$

and for our example, we have

$$s_{Y.X} = \sqrt{\frac{8.07}{10 - 2}} = 1.00$$

Exercises

2.1 Using the method of least squares, find the values of a and b in the equation $Y' = a + bX$ for the paired (X,Y) values in the following table. Draw the regression line of Y on X.

X	Values of Y
2	3, 6
4	2, 4, 8
6	5, 7, 10
8	5, 8, 10
10	8, 12
12	5, 9, 11

2.2 (a) Find the line of best fit for the equation $Y' = a + bX$ for the following paired (X,Y) values: $(6,6)$, $(5,4)$, $(4,5)$, $(3,3)$, $(2,1)$, $(1, -1)$, $(-1, -2)$, $(-2, -4)$, $(-3, -3)$, $(-4, -5)$. (b) Calculate the residual variance and the standard error of estimate. (c) What are the values of $\overline{Y}$ and $\overline{X}$? (d) What are the values of s_Y^2 and s_X^2?

2.3 Prove that $\Sigma(X - \overline{X})(Y - \overline{Y}) = \Sigma XY - (\Sigma X)(\Sigma Y)/n$.

2.4 Prove that $\Sigma(Y - Y')^2 = \Sigma y^2 - (\Sigma xy)^2/\Sigma x^2$.

2.5 If b is equal to zero, will c_{XY} be equal to zero?

2.6 We can write the following identity:

$$Y - \overline{Y} = (Y - Y') + (Y' - \overline{Y})$$

if we square and sum the above expression over all n values we have

$$\Sigma(Y - \overline{Y})^2 = \Sigma(Y - Y')^2 + \Sigma(Y' - \overline{Y})^2 + 2\Sigma(Y - Y')(Y' - \overline{Y})$$

Substitute an appropriate identity for Y' in the last term on the right and show that

$$2\Sigma(Y - Y')(Y' - \overline{Y}) = 0$$

2.7 Assume that

$$Y = a + bX + e$$

where $a = 5$ and $b = .5$. Assume also that we have $n = 6$ subjects and that for each subject we have some information available as shown below. Fill in the blank spaces in the table.

Subjects	Y'	X	Y	e
1		3		−1
2	25			2
3		5	30	
4		0		3
5		2	10	
6			11	

2.8 For which subject in the preceding exercise do we have the largest absolute value of $e = Y - Y'$? For which subject do we have the smallest absolute value of $e = Y - Y'$?

2.9 Briefly explain the meaning of each of the following concepts or terms:

linear relationship   regression coefficient
regression equation   covariance
residual variance   standard error of estimate
line of best fit   regression line
variance   perfect relationship
mean   standard deviation

Three

Correlation and Regression

3.1 Introduction

In our discussion in the first two chapters, it was assumed that we had some basis for designating one of the two variables investigated as the dependent variable Y and the other as the independent variable X. For example, if we were to measure vocabulary at selected age levels, it would seem logical to designate the vocabulary measure as the dependent variable and age as the independent variable. Vocabulary may depend on age, but it is rather difficult to imagine age as depending on vocabulary. Or suppose that one of our variables is the number of trials in a learning experiment and the other variable is the amount learned on each trial. Again it seems reasonable to regard the amount learned as depending on the number of trials rather than the number of trials as depending on the amount learned. If one of our variables is amount remembered and the other is time elapsed since learning, it seems logical to regard the amount remembered as depending on the passage of time rather than the other way around. In problems of the kind described, the experimenter ordinarily selects certain values of the independent variable X for investigation and subsequently observes the values of the dependent variable Y. The experimenter is interested in relating the values of the dependent variable Y to those of the independent variable X.

In many problems involving the relationship between two variables, however, there is no clear basis for designating one of the variables as the independent variable and the other as the dependent variable. If we have measured the heights of husbands and also the heights of their wives, which set of measures shall we designate as the dependent variable? If we have scores on a test of submissiveness and also on a test of aggressiveness, shall we consider the measure of submissiveness or the measure of aggressiveness as the dependent variable?

In problems of the kind just described, which variable we choose to designate as the dependent variable and which we choose to call the independent variable is a more or less arbitrary decision. If we arbitrarily designate one of the variables as Y and the other as X, then we may consider not only the prediction of Y values from the X values, but also the prediction of X values from the Y values. In other words, we may reverse the roles of our variables, first considering Y as a dependent variable with X as the independent variable, and then considering X as a dependent variable with Y as the independent variable.

In this chapter we shall assume there exists a population of paired (X,Y) values of the kind we have just described.[1] If we draw a sample from this population, neither the values of X nor the values of Y will be fixed. Instead, both the X values and the Y values will be subject to random variation. Under these circumstances we shall have not one but two regression lines. One will be

[1] As an example of a population of paired (X,Y) values, consider all male students enrolled at a given university as the population of interest, and let X be the height and Y be the weight of a student. For a sample of n students, we would have n ordered pairs of (X,Y) values.

for the regression of Y on X and the other will be for the regression of X on Y. Furthermore, we shall have two residual variances and two standard errors of estimate. There is, however, one statistic involving both variables for which we shall have but a single value. That statistic is the correlation coefficient.

3.2 The Correlation Coefficient

In discussing the correlation coefficient, we shall again restrict ourselves to the case of linear relationships. For convenience, we shall assume that one of our variables, Y, is a dependent variable and that the other variable, X, is an independent variable, so that we shall be concerned with the regression of Y on X. We will then reverse the roles of our variables and take X as the dependent variable and Y as the independent variable.

When the variables are linearly related, the *correlation coefficient* is a measure of the degree of relationship present. Consider first the existence of a perfect positive relationship between two variables, as shown in Figure 3.1(*a*).

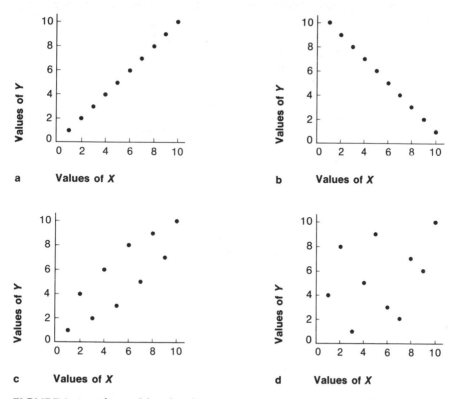

FIGURE 3.1 Plots of (X,Y) values for which: (*a*) $r = 1.00$, (*b*) $r = -1.00$, (*c*) $r = .84$, and (*d*) $r = .33$.

In this instance, as we shall see, the correlation coefficient is equal to 1.00. Figure 3.1(b) shows the plot of a set of X and Y values for which the correlation coefficient is equal to -1.00 and this is a perfect negative relationship. In Figure 3.1(c) the correlation between X and Y is equal to .84 and in Figure 3.1(d) the correlation coefficient is equal to .33.

3.3 Formulas for the Correlation Coefficient

The correlation coefficient may be defined as the covariance of X and Y divided by the product of the standard deviations of the X and Y variables, or

$$r = \frac{c_{XY}}{s_X s_Y} \tag{3.1}$$

We note also that

$$c_{XY} = r s_X s_Y \tag{3.2}$$

Substituting some identities in (3.1), we also have the following equations for the correlation coefficient:

$$r = \frac{\Sigma(X - \bar{X})(Y - \bar{Y})/(n - 1)}{\sqrt{\dfrac{\Sigma(X - \bar{X})^2}{n - 1}} \sqrt{\dfrac{\Sigma(Y - \bar{Y})^2}{n - 1}}} \tag{3.3a}$$

$$r = \frac{\Sigma XY - (\Sigma X)(\Sigma Y)/n}{\sqrt{\Sigma X^2 - (\Sigma X)^2/n} \sqrt{\Sigma Y^2 - (\Sigma Y)^2/n}} \tag{3.3b}$$

$$r = \frac{\Sigma xy}{\sqrt{\Sigma x^2} \sqrt{\Sigma y^2}} \tag{3.3c}$$

Table 3.1 shows a set of paired (X,Y) values. Substituting the appropriate values from Table 3.1 in (3.3b), we have

$$r = \frac{274 - (32)(56)/8}{\sqrt{156 - (32)^2/8} \sqrt{504 - (56)^2/8}}$$

$$= \frac{50}{\sqrt{28} \sqrt{112}}$$

$$= .89$$

3.4 The Regression of Y on X

We have previously shown that the regression coefficient is equal to the ratio of the covariance of the two variables to the variance of the independent

TABLE 3.1   A Sample of *n* = 8 Values of (*X*, *Y*)

	(1) X	(2) Y	(3) X^2	(4) Y^2	(5) XY	(6) Y'	(7) $Y - Y'$	(8) $(Y - Y')^2$
	7	13	49	169	91	12.37	.63	.3969
	6	9	36	81	54	10.58	−1.58	2.4964
	5	11	25	121	55	8.79	2.21	4.8841
	4	7	16	49	28	7.00	.00	.0000
	4	5	16	25	20	7.00	−2.00	4.0000
	3	7	9	49	21	5.21	1.79	3.2041
	2	1	4	1	2	3.42	−2.42	5.8564
	1	3	1	9	3	1.63	1.37	1.8769
Σ	32	56	156	504	274	56.00	0.00	22.7148

variable. Thus when Y was considered to be the dependent variable and X the independent variable, we had

$$b_Y = \frac{c_{XY}}{s_X^2} = \frac{\Sigma xy}{\Sigma x^2}$$

We now use the subscript Y to indicate that the regression coefficient is for Y on X. In our previous discussion of regression we were concerned only with the regression of Y on X and the subscript was not necessary.

We have already found that $\Sigma xy = 50$ and that $\Sigma x^2 = 28$ for the data in Table 3.1. The value of the regression coefficient b_Y is therefore

$$b_Y = \frac{50}{28} = 1.79$$

For the same data we have

$$\overline{Y} = \frac{56}{8} = 7$$

and

$$\overline{X} = \frac{32}{8} = 4$$

Then the regression equation for predicting Y values from the X values will be

$$Y' = a + b_Y X$$

where $a = \overline{Y} - b_Y \overline{X} = 7 - (1.79)(4) = -.16$. Substituting in the regression equation the values $a = -.16$ and $b_Y = 1.79$, we have, for the data in Table 3.1,

$$Y' = -.16 + 1.79X$$

3.5 The Residual Variance and Standard Error of Estimate

The residual sum of squares will be given by

$$\Sigma(Y - Y')^2 = \Sigma y^2 - \frac{(\Sigma xy)^2}{\Sigma x^2}$$

$$= 112 - \frac{(50)^2}{28}$$

$$= 22.71$$

which is equal, within rounding errors, to the value shown in Table 3.1. Dividing the residual sum of squares by $n - 2$, we obtain the residual variance

$$s_{Y.X}^2 = \frac{\Sigma(Y - Y')^2}{n - 2} = \frac{22.71}{6} = 3.7850$$

If we now take the square root of the residual variance, we obtain the standard error of estimate, or

$$s_{Y.X} = \sqrt{3.7850} = 1.9455$$

The standard error of estimate, $s_{Y.X}$, as we have pointed out previously, is a measure of the variability of the Y values about the regression line of Y on X. The standard deviation of the Y values, on the other hand, is a measure of the variation of the Y values about the mean of the Y values. In the present problem, the standard deviation of the Y values is

$$s_Y = \sqrt{\frac{112}{8 - 1}} = 4.0$$

3.6 The Regression of X on Y

If we now consider X as the dependent variable and Y as the independent variable, then the regression equation for predicting X values from Y values will be

$$X' = a + b_X Y \tag{3.4}$$

where

$$a = \overline{X} - b_X \overline{Y} \tag{3.5}$$

and

$$b_X = \frac{\Sigma xy}{\Sigma y^2} \tag{3.6}$$

For the data in Table 3.1 we have

$$b_X = \frac{50}{112} = .45$$

We also have $\overline{X} = 4$ and $\overline{Y} = 7$. Then

$$a = 4 - (.45)(7) = .85$$

and the regression equation for predicting X values from Y values will be

$$X' = .85 + .45Y$$

The residual sum of squares or sum of squared errors of prediction will be given by

$$\Sigma(X - X')^2 = \Sigma x^2 - \frac{(\Sigma xy)^2}{\Sigma y^2}$$

For the data in Table 3.1 we have

$$\Sigma(X - X')^2 = 28 - \frac{(50)^2}{112}$$

$$= 5.68$$

Then the residual variance will be given by

$$s_{X.Y}^2 = \frac{\Sigma(X - X')^2}{n - 2}$$

and, in our example, we have

$$s_{X.Y}^2 = \frac{5.68}{8 - 2} = .9467$$

and the standard error of estimate is thus equal to

$$s_{X.Y} = \sqrt{.9467} = .973$$

3.7 The Two Regression Lines

Figure 3.2, which is based on the data in Table 3.1, shows both the regression line of Y on X and the regression line of X on Y. The regression line of Y on X indicates the predicted change in Y as X varies. A change of one unit in X results in a predicted change of 1.79 units in Y, and this is the value of the regression coefficient b_Y. Similarly, the regression line of X on Y indicates that with a one-unit change in the value of Y, there is a corresponding predicted change of .45 units in X, and this is the value of the regression coefficient b_X.

As the correlation coefficient approaches either of its two limiting

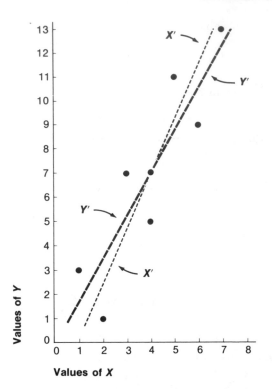

FIGURE 3.2 Plot of the (X, Y) values given in Table 3.1 showing the regression line (Y') of Y on X and the regression line (X') of X on Y.

values, -1.00 or 1.00, the two regression lines will move closer together. When r is equal to either -1.00 or 1.00, the two regression lines will coincide.

3.8 Correlation and Regression Coefficients

We see that if we consider the regression of X on Y instead of the regression of Y on X, we shall have corresponding formulas for the regression equation, regression coefficient, residual variance, and standard error of estimate. Although these formulas correspond in appearance, we should not expect them to result in identical numerical values. The only way in which these formulas could result in identical values would be if the means and the standard deviations of both the X and Y variables were identical. Let us see why this is so.

We have defined the regression coefficient of Y on X as

$$b_Y = \frac{\Sigma xy}{\Sigma x^2}$$

and we have also shown that

$$b_Y = \frac{c_{XY}}{s_X^2}$$

Recall also that

$$c_{XY} = r s_X s_Y$$

We thus have another commonly used expression for the regression coefficient of Y on X, or

$$b_Y = \frac{r s_X s_Y}{s_X^2} = r \frac{s_Y}{s_X} \tag{3.7}$$

The corresponding expression for the regression coefficient of X on Y can be obtained in the same manner and is

$$b_X = r \frac{s_X}{s_Y} \tag{3.8}$$

It is now readily apparent that only if the standard deviations of the X and Y variables are identical, will we also have $b_Y = b_X$ and, when this is true, both b_Y and b_X will be equal to r.

If we multiply the two regression coefficients, we obtain

$$b_X b_Y = r \frac{s_X}{s_Y} r \frac{s_Y}{s_X} = r^2$$

and

$$r = \pm \sqrt{r^2} = \pm \sqrt{b_X b_Y}$$

We note that both the regression coefficients must have the same sign and that the sign is determined by r. Consequently, if both b_Y and b_X are negative, then $\sqrt{r^2}$ is also negative. If both b_Y and b_X are positive, then $\sqrt{r^2}$ is positive.

3.9 Correlation and the Residual Sum of Squares

In the preceding chapter, we showed that $\Sigma(Y - Y')^2 = \Sigma y^2 - (\Sigma xy)^2/\Sigma x^2$, where $Y - Y'$ is an error of prediction resulting from the discrepancy between Y and Y' as predicted from the regression equation. If we multiply both the numerator and denominator of the last term on the right by Σy^2, we obtain

$$\Sigma(Y - Y')^2 = \Sigma y^2 - \frac{(\Sigma xy)^2 (\Sigma y^2)}{(\Sigma x^2)(\Sigma y^2)}$$

But

$$r^2 = \frac{(\Sigma xy)^2}{(\Sigma x^2)(\Sigma y^2)}$$

and substituting this identity in the preceding expression, we have

$$\Sigma(Y - Y')^2 = \Sigma y^2 - r^2 \Sigma y^2$$

or

$$\Sigma(Y - Y')^2 = \Sigma y^2 (1 - r^2) \tag{3.9}$$

The residual variance will then also be given by

$$s_{Y.X}^2 = \frac{\Sigma y^2 (1 - r^2)}{n - 2} \tag{3.10}$$

and the standard error of estimate will be given by

$$s_{Y.X} = \sqrt{\frac{\Sigma y^2 (1 - r^2)}{n - 2}} \tag{3.11}$$

A similar expression can be obtained for $\Sigma(X - X')^2$ and $s_{X.Y}^2$ by substituting Σx^2 for Σy^2 in (3.9). Thus, we also have

$$\Sigma(X - X')^2 = \Sigma x^2 (1 - r^2)$$

$$s_{X.Y}^2 = \frac{\Sigma x^2 (1 - r^2)}{n - 2}$$

$$s_{X.Y} = \sqrt{\frac{\Sigma x^2 (1 - r^2)}{n - 2}}$$

3.10 Correlation and the Regression Sum of Squares

In the preceding chapter we also showed that $\Sigma(Y' - \overline{Y})^2 = (\Sigma xy)^2/\Sigma x^2$. If we multiply both the numerator and the denominator of the term on the right by Σy^2, we have

$$\Sigma(Y' - \overline{Y})^2 = \frac{(\Sigma xy)^2 \Sigma y^2}{(\Sigma x^2)(\Sigma y^2)}$$

or

$$\Sigma(Y' - \overline{Y})^2 = r^2 \Sigma y^2 \tag{3.12}$$

Thus we see that the regression sum of squares is equal to $r^2 \Sigma y^2$.

3.11 Coefficients of Determination and Nondetermination

We have shown that the residual sum of squares, $\Sigma(Y - Y')^2$, can be expressed as $\Sigma y^2 (1 - r^2)$ and that the regression sum of squares, $\Sigma(Y' - \overline{Y})^2$, can be expressed as $r^2\Sigma y^2$. Consequently,

$$SS_{tot} = SS_{res} + SS_{reg}$$

can be expressed as

$$\Sigma y^2 = \Sigma y^2(1 - r^2) + r^2\Sigma y^2$$

Dividing both sides of the preceding expression by Σy^2, we have

$$1.00 = (1 - r^2) + r^2 \tag{3.13}$$

and r^2 is just the proportion of the total sum of squares, Σy^2, that can be accounted for by linear regression and $1 - r^2$ is the proportion that is independent of linear regression.

Now if we multiply both sides of (3.13) by s_Y^2, we have

$$s_Y^2 = s_Y^2(1 - r^2) + r^2 s_Y^2$$

Because $r_{XY} = r_{YX}$, we also have

$$s_X^2 = s_X^2(1 - r^2) + r^2 s_X^2$$

and we see that r^2 is the proportion of the Y variance that can be accounted for by the linear regression of Y on X and also the proportion of the X variance that can be accounted for by the linear regression of X on Y.

The value of r^2 is often referred to as the *coefficient of determination* and is commonly interpreted as the proportion of variance that X and Y have in common. The value of $1 - r^2$ is often referred to as the *coefficient of nondetermination* and is commonly interpreted as the proportion of variance that X and Y do not have in common.

Exercises

3.1 We have the following paired (X,Y) values:

X	Y
7	3
13	6
2	2
4	5
15	14
10	10
19	8
28	19
26	15
22	17

(a) Find the values of $\overline{X}$ and $\overline{Y}$. (b) Find the values of s_X^2 and s_Y^2. (c) Find the value of r. (d) Find the values of a and b in the regression equation $Y' = a + b_Y X$. (e) Find the values of a and b in the regression equation $X' = a + b_X Y$. (f) Find the values of $s_{Y.X}^2$ and $s_{X.Y}^2$.

3.2   (a) Find the correlation coefficient for the following paired (X,Y) values: $(12,12)$, $(10,13)$, $(9,9)$, $(8,8)$, $(7,5)$, $(6,6)$, $(4,0)$, $(2,2)$, $(1,1)$, $(0,3)$. (b) What are the values of b_Y and b_X?

3.3   Prove that $r^2 = b_X b_Y$.

3.4   Prove that the errors of prediction $Y - Y'$ are uncorrelated with the values of $X - \overline{X}$. *Hint:* It is sufficient to show that the numerator of the correlation coefficient is equal to zero.

3.5   Under what conditions will $r = b_Y$?

3.6   If $r = .60$, $s_Y = 8.0$, and $s_X = 10.0$, then what are the values of b_X and b_Y?

3.7   If $r = -1.00$, then what is the value of $\Sigma(Y - Y')^2$?

3.8   Is it possible for b_Y to be equal to 4.0 and for b_X to be equal to 2.0 for a given set of paired (X,Y) values? Explain why or why not.

3.9   If $b_Y = 2.0$, what is the maximum possible value for b_X?

3.10   If b_Y is equal to $-.60$, can b_X be equal to .60? Explain why or why not.

3.11   Under what conditions can b_X be equal to b_Y?

3.12   If the covariance c_{XY} is equal to r, then what do we know about s_Y and s_X?

3.13   Prove that $r_{Y'(Y-Y')} = 0$.

3.14   Prove that $r_{YY'} = r_{XY}$.

3.15   Prove that $r_{X(Y-Y')} = 0$.

3.16   Briefly explain the meaning of each of the following terms or concepts:

correlation coefficient
residual variance
coefficient of determination
coefficient of nondetermination

Four

Factors Influencing the Magnitude of the Correlation Coefficient

4.1 The Shapes of the *X* and *Y* Distributions

One of the factors influencing the magnitude of a correlation coefficient is the shape or form of the distribution of the *X* and *Y* values. Only if the distributions of the *X* and *Y* values are symmetrical and have the same shape or form is it possible to pair the *X* and *Y* values in such a way as to result in a correlation coefficient of either −1.00 or 1.00. Figure 4.1 shows three different symmetrical frequency distributions of *X* and *Y* values. In Figure 4.1(*a*) both the *X* and *Y* distributions are symmetrical and both are bell shaped. It is possible, therefore, to pair the *X* and *Y* values in such a way as to

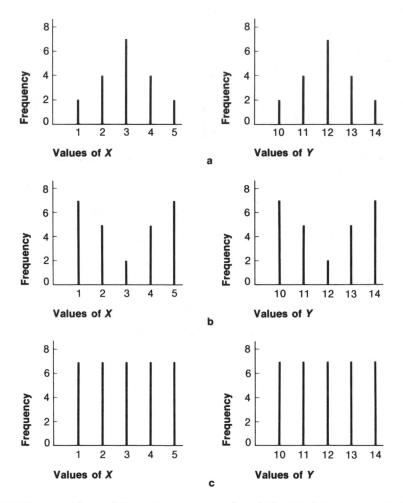

FIGURE 4.1   Three different symmetrical and identical frequency distributions for *X* and *Y*. The *X* and *Y* values for each frequency distribution can be paired so as to result in either *r* = 1.00 or *r* = −1.00.

result in a correlation coefficient of either −1.00 or 1.00. In Figure 4.1(*b*) both the *X* and *Y* distributions are symmetrical and both are U-shaped. Consequently, the *X* and *Y* values can be paired in such a way as to result in a correlation coefficient of either −1.00 or 1.00. In Figure 4.1(*c*) both the *X* and *Y* distributions are symmetrical and both are uniform or rectangular in shape. The *X* and *Y* values for these two distributions can also be paired so as to result in either *r* = −1.00 or *r* = 1.00.

In Figure 4.2(*a*) the frequency distributions of the *X* and *Y* values have the same shape or form, but the distributions are not symmetrical. Both distributions have the same degree of left skewness, that is, a tail to the left. In this example the *X* and *Y* values can be paired so as to result in a correlation coefficient equal to 1.00, but they cannot be paired in such a way as to result in a correlation coefficient of −1.00. Similarly, if the *X* and *Y* distributions have the same shape or form with the same degree of right skewness, that is, a tail to the right, then the values can be paired in such a way as to result in a correlation coefficient of 1.00 but not −1.00.

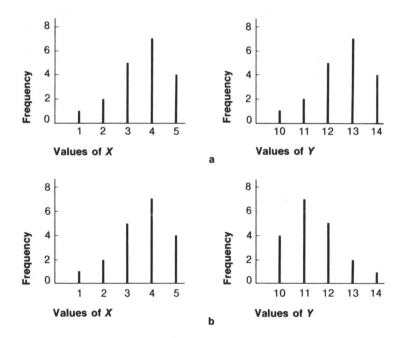

FIGURE 4.2   (*a*) A frequency distribution that has the same shape or form for both *X* and *Y* but that is not symmetrical. The *X* and *Y* values can be paired so as to result in *r* = 1.00 but not *r* = −1.00. (*b*) A frequency distribution for *X* that is left skewed and a frequency distribution for *Y* that has the same degree of right skewness. The *X* and *Y* values can be paired so as to result in *r* = −1.00 but not *r* = 1.00.

In Figure 4.2(*b*) the frequency distribution of the *X* values is left skewed and the frequency distribution of the *Y* values has the same degree of right skewness. In this example the *X* and *Y* values can be paired in such a way as to result in a correlation coefficient of -1.00, but they cannot be paired in such a way as to result in a value of 1.00.

4.2 Correlation Coefficients Based on Small Samples

A correlation coefficient based on a relatively small number of observations can be quite misleading. For example, if the population correlation coefficient is equal to zero and if random samples of $n = 5$ observations are drawn from the population, it can be expected that in 95 out of 100 samples the correlation coefficient will fall within the range $-.88$ to $.88$. Thus, a relatively high correlation coefficient of .70 would not be at all unusual in a sample of $n = 5$ observations, even though the population correlation coefficient is equal to zero.

With small samples a single pair of (X,Y) values may contribute excessively to the value of the correlation coefficient. Consider, for example, the paired (X,Y) values shown in Figure 4.3. If the point with coordinates (8,8) is included, the correlation coefficient is equal to .74. If the point with coordinates (8,8) is omitted, the correlation coefficient is equal to zero.

With small samples, we should always be conscious of the fact that an excessively high negative or positive value of the correlation coefficient may simply be the result of an extreme pair of (X,Y) values.

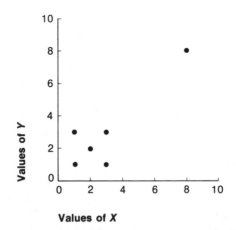

FIGURE 4.3   The correlation coefficient for the (X,Y) values is equal to .74. If the point with coordinates (8,8) is eliminated, the correlation coefficient is equal to zero.

4.3   Combining Several Different Samples

The magnitude of the correlation coefficient may also be influenced when a sample of observations really consists of two or more subsamples in which either the X means or the Y means, or both the X and Y means, differ from sample to sample. Before providing some examples of what may happen under these circumstances, we shall first review, briefly, the change in the two regression lines, Y on X and X on Y, as r increases from zero to one.

You will recall that when r is equal to zero, both b_X and b_Y are also equal to zero. Then the regression equation

$$Y' = \overline{Y} + b_Y(X - \overline{X})$$

results in a constant, $\overline{Y}$, for each value of X, and the "regression" line of Y on X is simply a horizontal line through the mean of the Y values. Similarly, if r is equal to zero, then the regression equation

$$X' = \overline{X} + b_X(Y - \overline{Y})$$

results in a constant, $\overline{X}$, for each value of Y, and the "regression" line of X on Y is simply a vertical line through the mean of the X distribution. These two lines are shown in Figure 4.4(a).

In Figure 4.4(b) we have $r = .20$. We have let $s_X = s_Y$ so that we also have $b_X = b_Y = r = .20$. Note that the two regression lines have moved slightly closer together. In Figure 4.4(c) we have $r = b_X = b_Y = .40$, and the two regression lines are closer together than they were when r was equal to .20. Figures 4.4(d) and 4.4(e) show the two regression lines when $r = b_X = b_Y = .60$ and when $r = b_X = b_Y = .80$, respectively. Note that the two regression lines move closer together as r increases. When $r = 1.00$, as shown in Figure 4.4(f), the two regression lines coincide. Similar changes in the two regression lines will occur with negative values of r, except that the two regression lines will then have negative slopes. When $r = -1.00$, the two regression lines will coincide.

Examine again Figure 4.4(a). With r equal to zero, the two regression lines are at right angles and can be enclosed in a circle. With r equal to .20, as shown in Figure 4.4(b), the two regression lines can be enclosed in a rather fat ellipse. Note that as r increases the ellipse enclosing the two regression lines becomes thinner and thinner. When r is equal to 1.00, the ellipse collapses into a single straight line.

In Figure 4.5(a) we have three samples. When each sample is considered separately, we have a fairly narrow ellipse, indicating that in each sample the correlation coefficient is relatively high. The ellipses slope upward from left to right, indicating that the correlation coefficients are positive. The X means for each sample are comparable but the Y means differ. If the three samples were combined we would have a relatively fat ellipse, and for the combined samples the correlation coefficient would be relatively low or close to zero.

The results just described may occur when a sample consists of both male

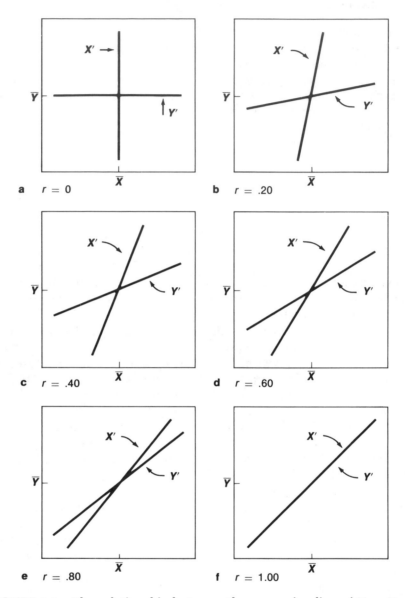

FIGURE 4.4 The relationship between the regression line of *Y* on *X* and the regression line of *X* on *Y* for various positive values of *r*. As *r* increases the two regression lines move closer together until, when *r* = 1.00, the two regression lines coincide.

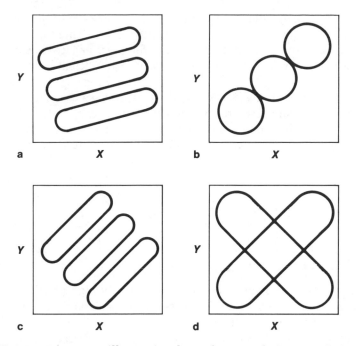

FIGURE 4.5   Diagrams illustrating how the correlation coefficient may be changed when different samples are combined. (*a*) In each sample the correlation coefficient is relatively high and positive. The correlation coefficient for the combined samples is close to zero. (*b*) In each sample the correlation coefficient is equal to zero. The correlation coefficient for the combined samples is moderately high and positive. (*c*) In each sample the correlation coefficient is high and positive. The correlation coefficient for the combined samples is low and negative. (*d*) In one sample the correlation coefficient is high and positive and in the other sample it is high and negative. The correlation coefficient for the combined samples is zero.

and female subjects. If there are systematic differences in either the X means or the Y means for the male and female subjects, then the correlation coefficient obtained by combining the male and female subjects into a single sample may differ considerably from the correlation coefficients obtained for the male and female subjects separately.

Figure 4.5(*b*) illustrates the results obtained for three groups or samples. When each sample is considered separately, the correlation coefficient is zero, as indicated by the three circles. If the three samples were combined, we would have a moderate ellipse and the correlation coefficient for the combined samples would be positive and moderately high. In this example the X and Y means are positively correlated.

In Figure 4.5(c) the correlation coefficients for each of the three groups are relatively high and positive, as indicated by the three ellipses. In this example, the X and Y means for the three groups are negatively correlated. If the three samples were combined, the correlation coefficient would be low and negative.

Caution must always be exercised in interpreting a correlation coefficient based on a composite of several groups when the X means and/or the Y means are different for the subgroups making up the composite.

In Figure 4.5(d) the two samples have approximately the same X and Y means. But in one of the samples the correlation coefficient is high and positive and in the other sample the correlation coefficient is high and negative. If the two samples were combined, the correlation coefficient for the combined samples would be zero.

4.4 Restriction of Range

Figure 4.6 shows two variables with a moderately high positive correlation coefficient. Suppose now that the range of the X variable in our sample is restricted. For example, suppose that the sample consists only of those individuals with X values falling to the right of the vertical line in the figure. The restriction in the range of the X variable would result in a lower correlation than would be obtained if the range were not so restricted.

As a not unrealistic example, assume that in a sample of unselected 7-year-old children, IQs and scores on a reading test are fairly highly positively correlated. A private school for gifted children, however, admits only those individuals aged 7 years with IQs equal to or greater than 130. This sample will have a restricted range of IQs and is also likely to have a restricted

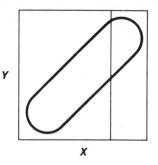

Y

X

FIGURE 4.6   In the population of unselected observations, the correlation coefficient is moderately high and positive. If a sample has a restricted range on the X variable consisting of only those observations with values falling to the right of the vertical line, the correlation coefficient will be considerably lower.

range of scores on the reading test. For this sample the correlation coefficient between IQs and scores on the reading test will be much lower than for the population of unselected 7-year-old children.

High school grades and first-year college grades of students in a given university may have a correlation of .60. If students who received low grades in high school are not admitted to the university, then the correlation coefficient of .60 is probably considerably lower than the correlation coefficient that would have been obtained if all students were admitted regardless of their high school grades.

If a correlation coefficient in a population is not equal to zero, then the value of the correlation coefficient for a sample in which the range of cases is restricted by selecting only those cases with values of X or Y falling at one extreme will tend to be closer to zero than the population correlation coefficient for unselected cases whose range is not restricted.

4.5 Nonlinearity of Regression

When X and Y have a curvilinear relationship rather than a linear relationship, the correlation coefficient may be quite low and, in fact, not applicable. Thus if a low value of a correlation coefficient is obtained, it may be worthwhile to examine a plot of the Y values against the X values in order to determine whether the Y values are related to the X values, but not linearly related.

4.6 Correlation with a Third Variable: Partial Correlation

A relatively high or low value of a correlation coefficient between two variables, X_1 and X_2, may be obtained depending on the correlation coefficients of the two variables with a third variable X_3. In some instances we may be interested in determining the value of the correlation coefficient between X_1 and X_2 when X_3 is held constant. Suppose, for example, we obtain the regression equation.

$$x_1' = b_1 x_3$$

where $b_1 = \Sigma x_1 x_3 / \Sigma x_3^2$ and $x_1' = X_1' - \overline{X}_1$. Then for each individual we could obtain the residual error

$$x_1 - x_1' = x_1 - b_1 x_3$$

These residual errors will be uncorrelated with X_3. For example,

$$\Sigma(x_1 - x_1')x_3 = \Sigma(x_1 - b_1 x_3)x_3$$
$$= \Sigma x_1 x_3 - b_1 \Sigma x_3^2$$
$$= 0$$

and consequently $r_{(x_1 - x_1')x_3}$ will be equal to zero. Similarly, we could obtain the regression equation

$$x_2' = b_2 x_3$$

where $b_2 = \Sigma x_2 x_3 / \Sigma x_3^2$ and $x_2' = X_2' - \overline{X}_2$. For each individual we could then obtain the residual error

$$x_2 - x_2' = x_2 - b_2 x_3$$

and these residuals could be shown in the same manner to be uncorrelated with X_3.

The correlation coefficient between the two sets of residuals represents the correlation between X_1 and X_2 with X_3 held constant. This correlation coefficient is called a *partial correlation coefficient*. The correlation coefficient between the two sets of residuals will be given by

$$r_{12.3} = \frac{r_{12} - r_{13} r_{23}}{\sqrt{1 - r_{13}^2} \sqrt{1 - r_{23}^2}}$$ (4.1)

where $r_{12.3}$ indicates the correlation between X_1 and X_2 with X_3 held constant.

The partial correlation coefficient $r_{12.3}$ will usually be smaller than the correlation coefficient r_{12}. For example, if we obtain the correlation coefficient between height and weight for children ranging in age from, say, 5 to 15 years, this correlation coefficient will be influenced by the fact that both height and weight are also positively correlated with age. Assume, for example, that X_1 represents height, X_2 represents weight, and X_3 represents age, and that the three correlation coefficients are $r_{12} = .90$, $r_{13} = .70$, and $r_{23} = .70$. Then the partial correlation coefficient between height and weight with age held constant will be

$$r_{12.3} = \frac{.90 - (.70)(.70)}{\sqrt{1 - .49} \sqrt{1 - .49}} = \frac{.41}{.51} = .80$$

As we have indicated, $r_{12.3}$ will, in general, tend to be smaller than r_{12}, but this is not necessarily always true. Suppose, for example, that with three variables either X_1 or X_2 has a zero correlation with X_3. Then, if $r_{12} = .80$, $r_{13} = .60$, and $r_{23} = 0$, we have

$$r_{12.3} = \frac{.80 - (.60)(.00)}{\sqrt{1 - .36} \sqrt{1}} = \frac{.80}{.80} = 1.00$$

In this interesting example, X_1 shares variance in common with X_3 but X_2 has no variance in common with X_3.

In essence, in this instance, the partial correlation coefficient removes the variance X_1 has in common with X_3 and the residuals are perfectly correlated with X_2. A variable X_3, which is correlated with X_1 or X_2 but has a zero correlation with the other variable, is commonly referred to as a *suppressor variable*. It suppresses the variance that either X_1 or X_2 has in common with X_3 but that is not part of the common variance of the other variable.

4.7 Random Errors of Measurement

Suppose that associated with each value of X_1 is a random error e_1 with mean equal to zero. Similarly, with each value of X_2 is associated a random error e_2 with mean equal to zero. We assume that the random errors are independent and also uncorrelated with the values of X_1 and X_2. In other words, if the values of X_1 and X_2 are put in deviation form, we assume that

$$\Sigma e_1 e_2 = \Sigma x_1 e_2 = \Sigma x_2 e_1 = \Sigma x_1 e_1 = \Sigma x_2 e_2 = 0 \qquad (4.2)$$

Then the correlation coefficient between X_1 and X_2 will be

$$r_{12} = \frac{\Sigma(x_1 + e_1)(x_2 + e_2)}{\sqrt{\Sigma(x_1 + e_1)^2}\sqrt{\Sigma(x_2 + e_2)^2}}$$

$$= \frac{\Sigma x_1 x_2 + \Sigma x_1 e_2 + \Sigma x_2 e_1 + \Sigma e_1 e_2}{\sqrt{\Sigma x_1^2 + 2\Sigma x_1 e_1 + \Sigma e_1^2}\sqrt{\Sigma x_2^2 + 2\Sigma x_2 e_2 + \Sigma e_2^2}}$$

or, because of (4.2),

$$r_{12} = \frac{\Sigma x_1 x_2}{\sqrt{\Sigma x_1^2 + \Sigma e_1^2}\sqrt{\Sigma x_2^2 + \Sigma e_2^2}}$$

We see that the denominator of the correlation coefficient between X_1 and X_2 includes the two terms Σe_1^2 and Σe_2^2. Then r_{12} will take its maximum value (positive or negative) only if $\Sigma e_1^2 = \Sigma e_2^2 = 0$, and this can occur only if all values of $e_1 = e_2 = 0$. Thus, the presence of random errors in the measurements of X_1 and X_2 will obviously result in a lower value for the correlation coefficient r_{12} than the value that would result in the absence of random errors in the measurements.

Measurements that are relatively free from random errors are commonly described as being reliable. There are obviously degrees of reliability, but the more reliable a set of measurements the smaller the error variance. If two psychological tests have relatively low reliabilities, that is, if each has a relatively large error variance, then the correlation between scores on the two tests will be considerably lower than the value that would be obtained if the two tests were more reliable.

Exercises

4.1   Under what conditions can X and Y values be paired in such a way as to result in either $r = 1.00$ or $r = -1.00$?

4.2   Assume that the correlation coefficient between X and Y for a sample of $n = 100$ males is equal to .60 and that, for a sample of $n = 100$ females, the correlation coefficient is equal to .70. The means of the X variable for both samples are approximately the same, but for the males $\overline{Y} = 60$ and for the females $\overline{Y} = 20$. If the two samples are combined, what would you predict about the correlation coefficient for the combined samples? Explain why.

4.3   The correlation coefficient in each of three independent samples is approximately equal to zero. The paired means $(\overline{X}, \overline{Y})$ for the three samples are (10,20), (20,30), and (30,40). What would you predict about the correlation coefficient for the combined samples? Explain why.

4.4   Assume that IQs and scores on a reading comprehension test are relatively highly correlated, $r = .80$, for all third-grade children in a given school district. If the correlation coefficient between the same two variables is obtained for only those children with IQs ranging from 90 to 110, what would you predict about this correlation coefficient? Explain why.

4.5   Suppose that the correlation coefficient r_{13} between X_1 and X_3 is equal to .70, and that the correlation coefficient r_{23} between X_2 and X_3 is equal to $-.70$. (a) Is it possible for the correlation coefficient r_{12} between X_1 and X_2 to be equal to .80? Explain why or why not. (b) Under the conditions described, what is the maximum possible positive value of r_{12}? (c) Under the conditions described, what is the maximum possible negative value of r_{12}?

4.6   Under what conditions might the partial correlation coefficient $r_{12.3}$ be higher than the correlation coefficient r_{12}?

4.7   For each of the examples shown below, predict what the correlation coefficient r_{XY} would be if the samples were combined.

(a)

Sample	$\overline{X}$	$\overline{Y}$	s_X	s_Y	r_{XY}
1	20	30	5	5	0
2	30	40	5	5	0
3	40	50	5	5	0

(b)

Sample	$\overline{X}$	$\overline{Y}$	s_X	s_Y	r_{XY}
1	20	25	5	5	0
2	25	20	5	5	0
3	30	15	5	5	0

(c)

Sample	$\overline{X}$	$\overline{Y}$	s_X	s_Y	r_{XY}
1	10	20	5	5	.80
2	20	20	5	5	.80
3	30	20	5	5	.80

(d)

Sample	$\overline{X}$	$\overline{Y}$	s_X	s_Y	r_{XY}
1	20	30	10	15	.70
2	20	30	10	15	.70

Five

Special Cases of the Correlation Coefficient

5.1 Introduction

There are some variables for which observations can take only one of two possible values. Variables of this kind are called *dichotomous*, *binomial*, or *binary* variables. A common example is the response to items on tests where the only possible responses are True or False or where the response is classified as correct or incorrect. Other examples of dichotomous variables are subjects classified as male or female, employed or unemployed, and married or single.

In some instances we may have two such dichotomous variables and we wish to determine whether there is any relationship between the two variables. For example, if we have two items in a test we may wish to determine whether responses to the two items are independent or whether they are related. In other instances, we may have one variable that is dichotomous and another variable for which the observations can take any one of a number of different values. As an example, we may have subjects classified as male and female and for each subject we also have a score on a test. We wish to determine whether there is any relationship between the sex classification of the subjects and their scores on the test.

5.2 The Phi Coefficient

We consider first a measure of the degree to which two dichotomous variables are related. Suppose, for example, we have $n = 200$ subjects who have responded to two True-False items. We arbitrarily designate response to one of the items as the X variable and response to the other item as the Y variable. We also arbitrarily assign a value of 1 to a True response and a value of 0 to a False response for each item. We thus have two dichotomous variables, X and Y, for which the observations can take a value of either 1 or 0.

With two dichotomous variables of the kind described, the possible paired (X,Y) values are limited to the following cases: $(1,0)$, $(1,1)$, $(0,0)$, and $(0,1)$, as shown in Figure 5.1(a). Note that for Item 1, the X variable,

$$\Sigma X = n_1$$

is simply the number of individuals who have given the 1 response to this item, and that

$$\overline{X} = \frac{n_1}{n} = p_1 \tag{5.1}$$

or the proportion of individuals who have given the 1 response to the item. We also have

$$\Sigma X^2 = n_1$$

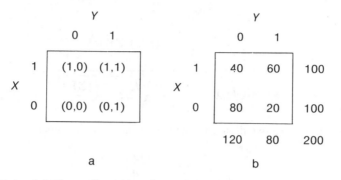

FIGURE 5.1   (a) The cell entries show the possible paired values of (X, Y) for two dichotomous variables. (b) The cell entries give the frequency of each of the paired (X, Y) values in a sample of $n = 200$.

and consequently

$$\Sigma(X - \bar{X})^2 = \Sigma X^2 - \frac{(\Sigma X)^2}{n}$$

$$= n_1 - \frac{n_1^2}{n} \tag{5.2}$$

Similarly, for response to Item 2, the Y variable, we have

$$\Sigma Y = n_2 \qquad \Sigma Y^2 = n_2 \qquad \bar{Y} = p_2$$

where n_2 is the number of individuals who have given the 1 response to Item 2 and p_2 is the proportion of individuals who have given the 1 response to Item 2. Then, we also have

$$\Sigma(Y - \bar{Y})^2 = \Sigma Y^2 - \frac{(\Sigma Y)^2}{n}$$

$$= n_2 - \frac{n_2^2}{n}$$

To determine the sum of the products of the paired (X, Y) values, we note that for the pairs $(1,0)$, $(0,0)$, and $(0,1)$, we have $XY = 0$, and that only the pairs $(1,1)$ will contribute to ΣXY. We let n_{12} be the number of individuals with the values $(1,1)$ and

$$\Sigma(X - \bar{X})(Y - \bar{Y}) = \Sigma XY - \frac{(\Sigma X)(\Sigma Y)}{n}$$

$$= n_{12} - \frac{n_1 n_2}{n} \tag{5.3}$$

We have

$$r = \frac{\Sigma XY - \dfrac{(\Sigma X)(\Sigma Y)}{n}}{\sqrt{\Sigma X^2 - \dfrac{(\Sigma X)^2}{n}}\sqrt{\Sigma Y^2 - \dfrac{(\Sigma Y)^2}{n}}} \tag{5.4}$$

and when there are two dichotomous variables, (5.4) becomes

$$r = \frac{n_{12} - \dfrac{n_1 n_2}{n}}{\sqrt{n_1 - \dfrac{n_1^2}{n}}\sqrt{n_2 - \dfrac{n_2^2}{n}}} \tag{5.5}$$

Dividing both the numerator and denominator of (5.5) by n, we have

$$r = \frac{p_{12} - p_1 p_2}{\sqrt{p_1 - p_1^2}\sqrt{p_2 - p_2^2}} \tag{5.6}$$

where p_{12} is the proportion of individuals who have given the 1 response to both Item 1 and Item 2. If we let $q_1 = 1 - p_1$ and $q_2 = 1 - p_2$, then

$$r = \frac{p_{12} - p_1 p_2}{\sqrt{p_1 q_1}\sqrt{p_2 q_2}} \tag{5.7}$$

The cell entries in Figure 5.1(b) correspond to the number of individuals with each of the paired (X, Y) values shown in the cells of Figure 5.1(a). For this example, we have

$$p_1 = \frac{100}{200} = .50$$

$$p_2 = \frac{80}{200} = .40$$

$$p_{12} = \frac{60}{200} = .30$$

Then, substituting in (5.7), we obtain

$$r = \frac{.30 - (.50)(.40)}{\sqrt{(.50)(.50)}\sqrt{(.40)(.60)}} = .41$$

The value of r we have just obtained is the correlation coefficient between two variables for each of which the observations can take only values of 1 or 0. This correlation coefficient is frequently referred to as a *phi coefficient*.

The calculation of the correlation coefficient between two dichotomous variables can be considerably simplified and we now show how this is possible. Note, however, that (5.7) for two dichotomous variables is just a special case of one of the standard formulas for calculating the correlation coefficient.

Examine Figure 5.2. The letters a, b, c, and d in the cells of Figure 5.2(b) now correspond to the number of individuals with each of the paired (X,Y) values shown in the cells of Figure 5.2(a). We observe that we now have the following identities:

$$\Sigma X = \Sigma X^2 = a + b$$
$$\Sigma Y = \Sigma Y^2 = b + d$$

and

$$\Sigma XY = b$$

We also have $n = a + b + c + d$. Substituting these identities in (5.4) we have

$$r = \frac{b - \dfrac{(a + b)(b + d)}{n}}{\sqrt{(a + b) - \dfrac{(a + b)^2}{n}}\sqrt{(b + d) - \dfrac{(b + d)^2}{n}}}$$

which can be shown to simplify to

$$r = \frac{bc - ad}{\sqrt{(a + c)(b + d)(a + b)(c + d)}} \tag{5.8}$$

	Y	
	0	1
1	(1,0)	(1,1)
0	(0,0)	(0,1)

X

a

	Y		
	0	1	
1	a	b	$a + b$
0	c	d	$c + d$
	$a + c$	$b + d$	n

X

b

FIGURE 5.2 (*a*) The cell entries show the possible paired values of (X,Y) for two dichotomous variables. (*b*) The cell entries correspond to the frequency of each of the paired (X,Y) values in a sample of *n* observations.

If we substitute in the preceding expression the corresponding entries from Figure 5.1(b), we have

$$r = \frac{(60)(80) - (40)(20)}{\sqrt{(120)(80)(100)(100)}} = .41$$

as before.

5.3 Range of r for Dichotomous Variables

In our earlier discussion of the correlation coefficient, we said that r has a possible range from -1.00 to 1.00 only if the frequency distributions of both X and Y are symmetrical and have the same form. For dichotomous variables, the frequency distribution is symmetrical only if $p_1 = q_1 = .50$, that is, when the proportion of individuals with $X = 1$ is equal to the proportion with $X = 0$. If for both the X and Y variables we have

$$p_1 = p_2 = .50$$

then it is possible for r to be equal to either -1.00 or 1.00. On the other hand, if

$$p_1 = p_2 \neq .50$$

then it is possible for r to be equal to 1.00, but it is not possible for r to be equal to -1.00. If

$$p_1 = q_2 \neq .50$$

then it is possible for r to be equal to -1.00, but it is not possible for r to be equal to 1.00.

5.4 The Point Biserial Coefficient

Suppose that one variable (X) is a dichotomous variable for which the observations can take only the values of 1 or 0, but that for the other variable (Y) the observations can take any one of a number of different possible values. Given that the X variable has possible values of 1 and 0, we note that for all paired values $(0,Y)$ the product XY will be equal to zero and that for all paired values $(1,Y)$ the product XY will be equal to Y. Then

$$\Sigma XY = \Sigma Y_1$$

or the sum of the Y values for those observations with an X value equal to 1.
For the X variable, we have

$$\Sigma X = \Sigma X^2 = n_1$$

or the number of observations with the value of $X = 1$. Consequently,

$$\Sigma(X - \overline{X})^2 = n_1 - \frac{n_1^2}{n}$$

Then for the correlation coefficient between a binary variable (X) and a variable that can take a number of different possible values (Y), we have, by substitution in (5.4),

$$r = \frac{\Sigma Y_1 - \dfrac{n_1 \Sigma Y}{n}}{\sqrt{n_1 - \dfrac{n_1^2}{n}} \sqrt{\Sigma Y^2 - \dfrac{(\Sigma Y)^2}{n}}} \tag{5.9}$$

For the data in Table 5.1 we have $\Sigma Y_1 = 80$, $\Sigma Y = \Sigma Y_0 + \Sigma Y_1 = 30 + 80 = 110$, and

$$\Sigma Y^2 = (4)^2 + (3)^2 + \cdots + (8)^2 = 770$$

Substituting these values as well as $n_1 = 10$ and $n = n_0 + n_1 = 10 + 10 = 20$ in (5.4), we obtain

$$r = \frac{80 - \dfrac{(10)(110)}{20}}{\sqrt{10 - \dfrac{(10)^2}{20}} \sqrt{770 - \dfrac{(110)^2}{20}}} = \frac{25}{\sqrt{5}\,\sqrt{165}} = \frac{5}{\sqrt{33}} = .87$$

TABLE 5.1   Values of $(0,Y)$ and $(1,Y)$ for a Dichotomous Variable X and a Variable Y That Can Take a Number of Different Values for a Sample of $n - 20$

	Values of Y	
	$(0,Y)$	$(1,Y)$
	4	8
	3	10
	3	10
	1	9
	2	6
	5	7
	5	7
	4	9
	1	6
	2	8
Σ	30	80

The correlation coefficient between a binary variable and a variable for which the observations can take any one of a number of different values is commonly referred to as the *point biserial coefficient.*

Note that the numerator of (5.9) can also be expressed as

$$\frac{n\Sigma Y_1 - n_1(\Sigma Y_0 + \Sigma Y_1)}{n} = \frac{1}{n}[\Sigma Y_1(n - n_1) - n_1\Sigma Y_0]$$

$$= \frac{1}{n}(n_0\Sigma Y_1 - n_1\Sigma Y_0)$$

and that the first term in the denominator of (5.9) can be expressed as

$$\sqrt{\frac{nn_1 - n_1^2}{n}} = \sqrt{\frac{n_1 n_0}{n}}$$

Substituting these expressions in (5.9), we have

$$r = \frac{\frac{1}{n}(n_0\Sigma Y_1 - n_1\Sigma Y_0)}{\sqrt{\frac{n_1 n_0}{n}}\sqrt{\Sigma Y^2 - \frac{(\Sigma Y)^2}{n}}} \tag{5.10}$$

Multiplying the numerator and the denominator of (5.10) by $n/n_1 n_0$, we obtain

$$r = \frac{\overline{Y}_1 - \overline{Y}_0}{\sqrt{\frac{n}{n_1 n_0}}\sqrt{\Sigma Y^2 - \frac{(\Sigma Y)^2}{n}}} \tag{5.11}$$

Then for the data in Table 5.1 we also have

$$r = \frac{8 - 3}{\sqrt{\frac{20}{(10)(10)}}\sqrt{770 - \frac{(110)^2}{20}}} = \frac{5}{\sqrt{.2}\sqrt{165}} = \frac{5}{\sqrt{33}} = .87$$

as before.

As (5.11) shows, if $\overline{Y}_1$ is equal to $\overline{Y}_0$, then the point biserial coefficient will be equal to zero. If X is an ordered variable, then the point biserial coefficient will be positive if $\overline{Y}_1 > \overline{Y}_0$ and negative if $\overline{Y}_1 < \overline{Y}_0$. If X is an unordered variable, then the sign of the point biserial coefficient is irrelevant.

It can be shown, by using some tedious algebra, that the square of (5.9), (5.10), or (5.11) is equal to

$$r^2 = \frac{n_0(\overline{Y}_0 - \overline{Y})^2 + n_1(\overline{Y}_1 - \overline{Y})^2}{\Sigma(Y - \overline{Y})^2} \tag{5.12}$$

For the data in Table 5.1, for example, we have

$$r^2 = \frac{(10)(3 - 5.5)^2 + 10(8 - 5.5)^2}{165}$$

$$= \frac{125}{165}$$

and, consequently,

$$r = \sqrt{\frac{125}{165}} = \sqrt{\frac{25}{33}} = \frac{5}{\sqrt{33}} = .87$$

as before.

In the analysis of variance, $\Sigma(Y - \overline{Y})^2$, based on two or more sets of Y values, is commonly referred to as the total sum of squares, or SS_{tot}. In the simplest case of the analysis of variance, SS_{tot} is partitioned into two independent components, the treatment sum of squares, SS_T, and the pooled within treatment sum of squares, SS_W. When we have two groups or sets of Y values, as we do for the point biserial coefficient, the treatment sum of squares is equal to the numerator of (5.12), that is,

$$SS_T = n_0(\overline{Y}_0 - \overline{Y})^2 + n_1(\overline{Y}_1 - \overline{Y})^2 \tag{5.13}$$

Then we also have as a formula for the point biserial coefficient,

$$r = \sqrt{\frac{SS_T}{SS_{tot}}} \tag{5.14}$$

Because

$$SS_{tot} = SS_T + SS_W \tag{5.15}$$

the pooled within treatment sum of squares can be obtained by subtraction. For the data in Table 5.1, for example, we have

$$SS_W = SS_{tot} - SS_T = 165 - 125 = 40$$

We can, of course, calculate SS_W. In our example it is equal to the sum of

$$\Sigma y_0^2 = \Sigma Y_0^2 - \frac{(\Sigma Y_0)^2}{n_0}$$

$$= 110 - \frac{(30)^2}{10}$$

$$= 20$$

and

$$\Sigma y_1^2 = \Sigma Y_1^2 - \frac{(\Sigma Y_1)^2}{n_1}$$

$$= 660 - \frac{(80)^2}{10}$$

$$= 20$$

or

$$SS_W = \Sigma y_0^2 + \Sigma y_1^2 = 20 + 20 = 40$$

It is obvious from (5.14) that the point biserial coefficient can be equal to 1.00 only if SS_T is equal to SS_{tot}, and, in this case SS_W would have to be equal to zero. But the only way in which SS_W can be equal to zero is if the n_0 values of $X = 0$ all have exactly the same Y_0 value and if the n_1 values of $X = 1$ also all have the same Y_1 value. In this instance, the point biserial coefficient would, in essence, be the same as the correlation coefficient between two binary variables for which r is equal to either -1.00 or 1.00.

In general, the Y values for the two categories of X will not have zero variance and, consequently, the point biserial coefficient will not be equal to 1.00. In the example shown in Table 5.1, we have maximized the value of the point biserial coefficient for the given values of Y. There is, in other words, no way in which these same Y values could be rearranged to obtain a value for the point biserial coefficient higher than .87.

5.5 The Rank Order Correlation Coefficient

Another special case of the correlation coefficient is when both the X and Y variables consist of a set of ranks. Suppose, for example, that two judges have ranked the same set of n objects according to some property of interest. We are interested in determining whether the ranks assigned to the objects by one judge are related to or show any agreement with the ranks assigned to the same objects by another judge.

If the values of a variable X consist of the ranks from 1 to n, then it can be shown that

$$\Sigma X = \frac{n(n + 1)}{2} \tag{5.16}$$

and that

$$\Sigma(X - \overline{X})^2 = \frac{n^3 - n}{12} \tag{5.17}$$

For the data in Table 5.2, we have

$$\Sigma X = \Sigma Y = \frac{10(10 + 1)}{2} = 55$$

and

$$\Sigma(X - \overline{X})^2 = \Sigma(Y - \overline{Y})^2 = \frac{10^3 - 10}{12} = 82.5$$

We also have

$$\Sigma XY = 378$$

Substituting these values in (5.4), we obtain

$$r = \frac{378 - \dfrac{(55)(55)}{10}}{\sqrt{82.5}\ \sqrt{82.5}} = \frac{75.5}{82.5} = .915$$

The correlation coefficient between two sets of ranks is commonly referred to as the *rank order correlation coefficient*. The calculation of the rank order correlation coefficient can be considerably simplified. For example, if for *any* pair of (X,Y) values we let

$$D = X - Y$$

then

$$\Sigma D - \Sigma X \quad \Sigma Y$$

TABLE 5.2   Rank Values of (X,Y) in a Sample of $n = 10$

	X	Y	XY	D = X Y	D²
	10	8	80	2	4
	9	10	90	−1	1
	8	9	72	−1	1
	7	7	49	0	0
	6	4	24	2	4
	5	6	30	−1	1
	4	5	20	−1	1
	3	3	9	0	0
	2	1	2	1	1
	1	2	2	−1	1
Σ	55	55	378	0	14

and

$$\overline{D} = \overline{X} - \overline{Y}$$

The sum of the squared deviations of the D values from the mean of the D values will be

$$\Sigma(D - \overline{D})^2 = \Sigma[(X - \overline{X}) - (Y - \overline{Y})]^2$$

$$= \Sigma x^2 + \Sigma y^2 - 2\Sigma xy$$

$$= \Sigma x^2 + \Sigma y^2 - 2\Sigma xy \frac{\sqrt{(\Sigma x^2)(\Sigma y^2)}}{\sqrt{(\Sigma x^2)(\Sigma y^2)}}$$

$$= \Sigma x^2 + \Sigma y^2 - 2r \sqrt{(\Sigma x^2)(\Sigma y^2)}$$

and

$$r = \frac{\Sigma x^2 + \Sigma y^2 - \Sigma(D - \overline{D})^2}{2\sqrt{(\Sigma x^2)(\Sigma y^2)}} \tag{5.18}$$

Substituting in the preceding expression the corresponding values for ranks, we obtain

$$r = \frac{\dfrac{n^3 - n}{12} + \dfrac{n^3 - n}{12} - \Sigma(D - \overline{D})^2}{2\sqrt{\left(\dfrac{n^3 - n}{12}\right)\left(\dfrac{n^3 - n}{12}\right)}}$$

or

$$r = 1 - \frac{6\Sigma(D - \overline{D})^2}{n^3 - n} \tag{5.19}$$

But if X and Y are ranks, then $\overline{X} = \overline{Y}$ and

$$\overline{D} = \overline{X} - \overline{Y} = 0$$

and, consequently,

$$r = 1 - \frac{6\Sigma D^2}{n^3 - n} \tag{5.20}$$

For the data in Table 5.2 we have

$$r = 1 - \frac{(6)(14)}{10^3 - 10} = .915$$

which is the same value we obtained before.

5.6 The Variance and Standard Deviation of the Difference between Two Independent Variables

We have shown that for *any* pair of (X,Y) values, if $D = X - Y$, then

$$\Sigma(D - \overline{D})^2 = \Sigma x^2 + \Sigma y^2 - 2\Sigma xy$$

Dividing both sides of this expression by $n - 1$, we have the variance of the differences or

$$s_D^2 = s_{X-Y}^2 = s_X^2 + s_Y^2 - 2c_{XY}$$

It can be shown that if X and Y are independent variables, then c_{XY} will be equal to zero. Consequently, if X and Y are independent variables, then the variance of the difference will be

$$s_{X-Y}^2 = s_X^2 + s_Y^2$$

and the standard deviation will be

$$s_{X-Y} = \sqrt{s_X^2 + s_Y^2} \tag{5.21}$$

Exercises

5.1 We have two dichotomous variables, X_1 and X_2. The numbers of observations in a sample of $n = 200$ with the response patterns $(1,0)$, $(1,1)$, $(0,0)$, and $(0,1)$ are 45, 45, 80, and 30, respectively. Find the value of the correlation coefficient between X_1 and X_2.

5.2 We know the scores of eight males and eight females on a test Y. We arbitrarily assign a value of $X = 0$ to the males and $X = 1$ to the females. The scores of the subjects on the test Y are as follows:

Males (X_0)	Females (X_1)
7	10
6	9
5	8
4	7
4	7
3	6
2	5
1	4

Find the value of the correlation coefficient between X and Y.

5.3 A group of male students and a group of female students rated eight personality traits in terms of their desirability in a marital partner. The average ratings for each trait for each group of subjects were then translated into ranks. The ranks assigned to the traits by the males and females are as follows:

Trait	Male	Female
A	1	7
B	2	3
C	3	1
D	4	5
E	5	8
F	6	4
G	7	6
H	8	2

Calculate the correlation coefficient between the two sets of ranks.

5.4   Two judges tasted and ranked each of nine domestic brands of red wine in terms of overall merit. The ranks assigned to the brands by each judge are as follows:

Brand	Judge 1	Judge 2
A	1	6
B	3	5
C	2	2
D	6	1
E	4	8
F	5	3
G	8	7
H	9	4
I	7	9

Calculate the correlation coefficient between the two sets of ranks.

5.5   A study of 100 women who thought their marriage was successful and 100 women who thought their marriage was unsuccessful revealed a differential in response to the question: Did you have a happy childhood? For these two dichotomous variables, the following frequencies were obtained:

Childhood status	Marital status	
	Unsuccessful	Successful
Happy	40	70
Unhappy	60	30

Find the value of the correlation coefficient between the two dichotomous variables.

5.6 Under what conditions can the phi coefficient be equal to either 1.00 or −1.00?

5.7 Under what conditions can the phi coefficient be equal to 1.00 but not −1.00?

5.8 Under what conditions can the phi coefficient be equal to −1.00 but not 1.00?

5.9 If $\overline{Y}_0 = \overline{Y}_1$, then what do we know about the value of the point biserial coefficient?

5.10 Under what conditions can the point biserial coefficient be equal to 1.00?

5.11 If $p_1 = .8$ and $p_2 = .2$, what is the maximum positive value of r? What is the maximum negative value of r?

5.12 If $p_1 = .5$ and $p_2 = .8$, what is the maximum positive value of r? What is the maximum negative value of r?

Six

Tests of Significance for Correlation Coefficients

6.1 Introduction

The calculation of the correlation coefficient r for a sample of n observations requires no assumptions about the distributions of either the X or Y variables, although the nature of these distributions may place limitations on the magnitude of the correlation coefficient in the manner in which we have indicated previously. If, however, we are interested in using the sample value of r to infer something about the corresponding population parameter ρ, then we do need to be concerned about distribution assumptions.

In a typical correlation problem, we assume that we have a random sample of n paired (X,Y) values. Note that the values of X are not preselected as in a typical regression problem. On the basis of the obtained sample correlation coefficient r, we wish to test a null hypothesis regarding the population value ρ. In order to use the available tests of significance, we must make certain assumptions regarding the distribution of both the X and the Y values. First of all, it is assumed that both the X and Y variables are normally distributed in the population with corresponding means μ_X and μ_Y and variances σ_X^2 and σ_Y^2. It is not necessary that $\mu_X = \mu_Y$ or that $\sigma_X^2 = \sigma_Y^2$. For each possible value of X there is a corresponding population of normally distributed Y values with mean $\mu_{Y.X}$ and variance $\sigma_{Y.X}^2$. The variance $\sigma_{Y.X}^2$ is assumed to be the same for each value of X. Similarly, for each possible value of Y there is a corresponding population of X values with mean $\mu_{X.Y}$ and with variance $\sigma_{X.Y}^2$, which is assumed to be the same for each value of Y. When these assumptions are met, the joint distribution of the paired (X,Y) values is said to be a *bivariate normal distribution.*

In general, tests of significance of correlation coefficients are based on the assumption that X and Y have a bivariate normal distribution.[1] With a bivariate normal distribution, the only possible relationship between X and Y is a linear relationship. It does not follow, however, that X and Y must be linearly related; that is, it is possible for ρ to be equal to zero, even though the population distribution is bivariate normal.

6.2 Tests of Significance

Ordinarily, in testing a null hypothesis for significance we decide in advance on some small probability, called the *significance level* of the test and represented by α, as a standard for deciding whether to reject the null hypothesis. Frequently used standards are $\alpha = .05$ and $\alpha = .01$, but other values of α might also be chosen. If the result of the test of significance is such that the probability of the outcome is equal to or less than α, then the null hypothesis is rejected. This simply means that the probability of the outcome

---

[1]An exception, as we shall show in Chapter 8, is the test of the null hypothesis that $\rho = 0$.

is sufficiently small that we choose to regard the null hypothesis as improbable or false.

A Type I error occurs when a null hypothesis is, in fact, true, but the test of significance results in a decision to reject it.[2] Our protection against making a Type I error is equal to $1 - \alpha$. For example, if we set $\alpha = .01$, this is the probability that we will make a Type I error, given that the null hypothesis is true; and our protection against making a Type I error, given that the null hypothesis is true, is equal to $1 - \alpha = .99$.

Tests of significance are commonly made by transforming a statistic or statistics based on a sample into another statistic for which the probability distribution is known, given that a null hypothesis is true. In this chapter, we make use of three different probability distributions: the standard normal distribution Z, the t distribution, and the χ^2 distribution. In Chapter 7 we make use of another probability distribution, the F distribution.

Each of these distributions is represented by a curve such that the area under the curve is equal to one. Tables in the Appendix give the proportion of the total area in the right tail of the curve when ordinates are erected at various tabled values of Z, t, χ^2, and F. These areas correspond to the probabilities of obtaining Z, t, χ^2, or F equal to or greater than the tabled value, when a null hypothesis is true, and can be used to test various null hypotheses.

There is only one standard normal distribution and, consequently, the table of the standard normal distribution is entered only in terms of Z. The t, χ^2, and F distributions depend on the number of degrees of freedom involved. Consequently, in using the t, χ^2, and F tables we need to know not only the observed value of t, χ^2, or F, but also the number of degrees of freedom associated with the observed value.

6.3 Sampling Distribution of the Correlation Coefficient

Suppose that there exists a population of paired (X, Y) values such that for this population the correlation coefficient is equal to ρ. If random samples of n are drawn from this population, each of the n observations will consist of an ordered pair of (X, Y) values, and for each random sample it will be possible to calculate the sample correlation coefficient r. If ρ is close to zero, then the sampling distribution of the r values will be approximately normal in form, provided n is not too small. However, if ρ is not close to zero and if n is small, then the sampling distribution of r will not be normal in form but will instead be skewed. For example, if $\rho = 0.80$ and $n = 8$, the sample values will tend to

---

[2]We should also be concerned about Type II errors. A Type II error occurs when a null hypothesis is false but the test of significance fails to reject it. Other things being equal, the probability of a Type II error decreases as the sample size increases.

cluster around 0.80, but there will be a tail to the left; that is, the distribution will be left skewed.

6.4 Test of the Null Hypothesis that $\rho = 0$

Table 6.1 shows the X and Y values for two independent random samples. For the first sample we have $n = 15$ observations and for the second sample we have $n = 10$ observations. For the first sample, we have

$$r = \frac{564 - (63)(110)/15}{\sqrt{365 - (63)^2/15}\ \sqrt{934 - (110)^2/15}} = \frac{102}{\sqrt{100.40}\ \sqrt{127.33}} = .902$$

To test the null hypothesis that ρ, the population correlation coefficient, is equal to zero, we calculate

$$t = \frac{r}{\sqrt{1 - r^2}}\ \sqrt{n - 2} \tag{6.1}$$

Then t, as defined by (6.1), will be distributed in accordance with the tabled values of t with degrees of freedom equal to $n - 2$, when the null hypothesis is true. For example, with $15 - 2 = 13$ d.f., we find from the table of t, Table III in the Appendix, that the probability of obtaining $t \geq 3.012$ is .005, if it is true that $\rho = 0$. Because the distribution of t is symmetrical, when the null hypothesis is true, the probability of obtaining $t \leq -3.012$ is also .005. Then the probability of obtaining $t \geq 3.012$ or $t \leq -3.012$ is equal to $.005 + .005 = .01$, if $\rho = 0$.

Let us assume that we have chosen $\alpha = .01$. Because we are interested in the possibility that ρ may be either positive or negative, if it is not equal to zero, we will reject the null hypothesis if either $t \geq 3.012$ or $t \leq -3.012$. For our example, we have

$$t = \frac{.902}{\sqrt{1 - (.902)^2}}\ \sqrt{15 - 2} = 7.53$$

Because our obtained $t = 7.53$ exceeds the tabled value 3.012, we reject the null hypothesis.[3]

6.5 Table of Significant Values of r

It is possible to substitute in (6.1) the various values of n and the tabled values of t and to solve for the values of r that would be required at various levels of

[3]From the table of t, Table III in the Appendix, we find that the probability of obtaining t equal to or greater than 7.53, when the null hypothesis is true, is approximately .0000025. Consequently, the probability of obtaining $t \geq 7.53$ or $t \leq -7.53$ is approximately .000005.

TABLE 6.1 Values of (X, Y) for Two Independent Random Samples of n = 15 and n = 10

	X	Y	X²	Y²	XY
			Sample 1: *n* = 15		
	6	10	36	100	60
	1	6	1	36	6
	1	3	1	9	3
	1	4	1	16	4
	6	9	36	81	54
	7	10	49	100	70
	2	3	4	9	6
	4	8	16	64	32
	8	11	64	121	88
	8	11	64	121	88
	1	6	1	36	6
	5	10	25	100	50
	7	10	49	100	70
	3	5	9	25	15
	3	4	9	16	12
Σ	63	110	365	934	564
			Sample 2: *n* = 10		
	X	Y	X²	Y²	XY
	3	2	9	4	6
	1	4	1	16	4
	1	2	1	4	2
	7	7	49	49	49
	6	6	36	36	36
	3	2	9	4	6
	1	4	1	16	4
	5	6	25	36	30
	8	10	64	100	80
	7	8	49	64	56
Σ	42	51	244	329	273

significance. This has been done and the resulting values of *r* to three decimal places are given in Table IV in the Appendix. The values of *r* given in Table IV are those that would be regarded as significant with probabilities given by the column headings if one-sided tests are made using the right tail of the *t* distribution, that is, for positive values of *r*. Because the distribution of *r* is symmetrical when $\rho = 0$, the table can also be used with negative values of r. For example, with $n - 2 = 13$ d.f., we find that the probability of obtaining $r \geq .641$ is $.005$ when $\rho = 0$. This is also the probability of obtaining $r \leq -.641$ when $\rho = 0$. For a two-sided test of the null hypothesis, the probability of obtaining either $r \geq .641$ or $r \leq -.641$ is $.01$.

It is evident from Table IV that a relatively large observed value of r may not be significant when r is based on a small number of observations. On the other hand, as n increases, relatively small values of r will result in the rejection of the null hypothesis. For example, if a correlation coefficient is based on $n = 1000$ observations, then values of r equal to or greater than $.06$ or equal to or less than $-.06$ have a probability of approximately $.05$, when the null hypothesis is true.

6.6 The z_r Transformation for r

Any value of r may be transformed to a new variable z_r, defined as

$$z_r = \frac{1}{2} \left[\log_e(1 + r) - \log_e(1 - r) \right] \tag{6.2}$$

where r is the observed value of the correlation coefficient. In order to make the z_r values of (6.2) available independently of a table of natural logarithms, values of r were substituted in (6.2) and the corresponding values of z_r were obtained. These z_r values are given in Table V in the Appendix. In our example, we have $r = .902$, and from Table V we find that the corresponding value of z_r is approximately 1.472. The z_r distribution is symmetrical about zero, and if a correlation coefficient has a minus sign, then so also will the corresponding value of z_r. For example, if $r = -.902$, then $z_r = -1.472$.

Fisher[4] has shown that the distribution of z_r is approximately normal in form and that for all practical purposes the distribution is independent of the population value ρ. This means that the distribution of z_r remains approximately normal in form even when samples are drawn from populations in which ρ is large. Furthermore, the standard error of z_r is related in a very simple way to n, the sample size, and is given by

$$\sigma_{z_r} = \frac{1}{\sqrt{n - 3}} \tag{6.3}$$

where n is the number of observations on which r is based. For the first sample in Table 6.1 we have $n = 15$ observations and

$$\sigma_{z_r} = \frac{1}{\sqrt{15 - 3}} = .29$$

---

[4]Fisher, R. A. On the "probable error" of a coefficient of correlation deduced from a small sample. *Metron*, 1921, *1*, Part 4, 1–32.

If z_r is approximately normally distributed, then

$$Z = \frac{z_r - z_\rho}{\sigma_{z_r}} \tag{6.4}$$

will have a distribution that is approximately that of a standard normal variable with $\mu = 0$ and $\sigma = 1$ and can be evaluated in terms of the table of the standard normal distribution, Table I in the Appendix. In (6.4), z_r is the value corresponding to the observed value of r and z_ρ is the value corresponding to ρ.

Suppose we want to make a two-sided test of the null hypothesis that $\rho = .60$, with $\alpha = .01$. From Table V we find that $z_\rho = .693$. From the table of the standard normal distribution, Table I, we find that the probability of obtaining $Z \geq 2.58$, when the null hypothesis is true, is .005, and this is also the probability of obtaining $Z \leq -2.58$. Thus, with a two-sided test, we will reject the null hypothesis if we obtain Z equal to or greater than 2.58 or Z equal to or less than -2.58. Substituting the values $z_r = 1.472$, $z_\rho = .693$, and $\sigma_{z_r} = .29$ in (6.4), we have

$$Z = \frac{1.472 - .693}{.29} = 2.686$$

and because Z is greater than 2.58, the null hypothesis is rejected. We conclude that it is improbable that the population correlation is as low as .60.

6.7 Establishing a Confidence Interval for ρ

Rather than testing various null hypotheses regarding the value of ρ, suppose that we set up the following inequality:

$$-2.58 < \frac{z_r - z_\rho}{\sigma_{z_r}} < 2.58$$

If $z_r = 1.472$ and $\sigma_{z_r} = .29$, we have

$$-2.58 < \frac{1.472 - z_\rho}{.29} < 2.58$$

or

$$1.472 + (.29)(2.58) > z_\rho > 1.472 - (.29)(2.58)$$

$$2.220 > z_\rho > .724$$

From the table of z, we find that the two r's corresponding to $z_r = 2.220$ and $z_r = .724$ are approximately .975 and .620, respectively.

The interval, .620 to .975, that we have just established is called a 99 percent *confidence interval* for ρ. The value .620 is called the *lower confidence limit* and the value .975 is called the *upper confidence limit*. Confidence limits are statistics and, like all statistics, may be expected to vary from sample to sample. If we have a new and independent sample of $n = 15$ from the same population as the present one, we will not necessarily find that r for this sample is equal to .902 and, consequently, the lower and upper 99 percent confidence limits will not be the same as those obtained with the present sample. We can, however, make the statement that we are 99 percent confident that a 99 percent confidence interval will contain ρ. The basis for this statement is that if we have an indefinitely large number of samples, we can expect that in 99 out of 100 samples, the 99 percent confidence interval will contain ρ and that in 1 out of 100 samples, ρ will not be contained within the interval. We, of course, have no way of knowing whether the obtained interval is one of the 99 in 100 that contain ρ or whether it is the 1 in 100 that does not. However, if we always infer that ρ falls within a 99 percent confidence interval, we can expect that, in the long run, 99 percent of our inferences will be correct and only 1 percent will be incorrect.

We note that the confidence limits, .620 and .975, are not equally distant from the observed value of $r = .902$. The lower confidence limit deviates .282 from the observed value of r and the upper confidence limit deviates .073 from r. However, as n increases, the lower and upper confidence limits on the r scale will become more symmetrical about the observed value of r. For example, if $r = .902$ were based on a sample of $n = 403$ observations, then $\sigma_{z_r} = 1/\sqrt{403 - 3} = .05$. If we now find a 99 percent confidence interval for ρ, we will observe that the lower limit on the r scale is approximately .87 and the upper limit is approximately .92. These two values, based on a sample of $n = 403$ observations, are more symmetrical about the observed value of $r = .902$ than are the two values based on a sample of $n = 15$ observations. What this indicates, of course, is that as n increases, the skewness of the sampling distribution of r decreases.

6.8 Test of Significance of the Difference between r_1 and r_2

One of the advantages of the z_r transformation for r is that is also permits us to test the significance of the difference between two values of r obtained from two independent samples. To illustrate the test of significance, we use the data given in Table 6.1 for two samples. We have already found that for the first sample of $n = 15$ observations, $r_1 = .902$. For the second sample, we have

$$r_2 = \frac{273 - (42)(51)/10}{\sqrt{244 - (42)^2/10}\sqrt{329 - (51)^2/10}} = \frac{58.8}{\sqrt{(67.6)(68.9)}} = .862$$

The two values $r_1 = .902$ and $r_2 = .862$ are not equal. Is it reasonable to believe that the two samples are from a common population so that $\rho_1 = \rho_2$? If we reject this null hypothesis, we will conclude that $\rho_1 \neq \rho_2$; in other words, that the difference between $r_1 = .902$ and $r_2 = .862$ is sufficiently large that we do not believe they are both estimates of the same population value ρ.

To make the test of significance, we transform both r_1 and r_2 into z_r values. The standard error of the difference between two independent values of z_r will be given by the usual formula for the standard error of the difference between two independent variables[5] or, in the case of two z_r values, z_1 and z_2,

$$\sigma_{z_1 - z_2} = \sqrt{\sigma_{z_1}^2 + \sigma_{z_2}^2}$$

$$= \sqrt{\frac{1}{n_1 - 3} + \frac{1}{n_2 - 3}} \tag{6.5}$$

In our example, we have $n_1 = 15$ and $n_2 = 10$, and therefore

$$\sigma_{z_1 - z_2} = \sqrt{\frac{1}{15 - 3} + \frac{1}{10 - 3}} = .476$$

Then, the difference between z_1 and z_2 divided by the standard error of the difference results in

$$Z = \frac{z_1 - z_2}{\sigma_{z_1 - z_2}} \tag{6.6}$$

If the null hypothesis $\rho_1 = \rho_2$ is true, then Z will have a distribution that is approximately that of a standard normal variable with $\mu = 0$ and $\sigma = 1$ and can be evaluated in terms of the table of the standard normal distribution, Table I in the Appendix.

In our example, we have $r_1 = .902$ with $z_1 = 1.472$ and $r_2 = .862$ with $z_2 = 1.293$. Then, substituting these two values of z_r and $\sigma_{z_1 - z_2} = .476$ in (6.6), we have

$$Z = \frac{1.472 - 1.293}{.476} = .376$$

For a two-sided test, using the table of the standard normal distribution, we find that the probability of $Z \geq .376$ or $Z \leq -.376$ is about .71, when the null hypothesis $\rho_1 = \rho_2$ is true. We may regard the null hypothesis as tenable and conclude that the difference between the two correlation coefficients is not sufficiently great to cause us to believe that they are not both estimates of the same population value ρ.

Because the two values of r can be considered estimates of the same population value ρ, we calculate

$$\bar{z}_r = \frac{(n_1 - 3)z_1 + (n_2 - 3)z_2}{(n_1 - 3) + (n_2 - 3)} \tag{6.7}$$

[5]See Section 5.6.

which is the weighted average value of z_r. The weighted average in the present problem is

$$\bar{z}_r = \frac{(15 - 3)(1.472) + (10 - 3)(1.293)}{(15 - 3) + (10 - 3)} = \frac{26.715}{19} = 1.406$$

From Table V we find that the corresponding value of r is approximately .885. We may regard this value as an estimate of the common population value ρ based on the data from the two samples.

6.9 The χ^2 Test for the Difference between r_1 and r_2

We now consider an equivalent method for testing the null hypothesis $\rho_1 = \rho_2$ for two independent random samples using the χ^2 distribution. This method is of value in that it can also be used to test the homogeneity of more than two sample values of r. For example, if we have several independent random samples, each with a given value of r, we may wish to test the null hypothesis $\rho_1 = \rho_2 = \cdots = \rho_k$. We can make this test in the case of $k = 2$ samples by finding

$$\chi^2 = (n_1 - 3)z_1^2 + (n_2 - 3)z_2^2 - \frac{[(n_1 - 3)z_1 + (n_2 - 3)z_2]^2}{(n_1 - 3) + (n_2 - 3)} \quad (6.8)$$

with $k - 1 = 1$ d.f. For our two samples, we have

$$\chi^2 = (15 - 3)(1.472)^2 + (10 - 3)(1.293)^2$$

$$- \frac{[(15 - 3)(1.472) + (10 - 3)(1.293)]^2}{(15 - 3) + (10 - 3)}$$

$$= 37.704 - \frac{(26.715)^2}{19}$$

$$= .141$$

and we find, from the table of χ^2, Table II in the Appendix, that with 1 d.f., the probability of obtaining $\chi^2 \geq .141$, when the null hypothesis is true, is approximately .70.

Because Z as defined by (6.6) is a standard normal variable, and because the square of a standard normal variable is a value of χ^2 with 1 d.f., the χ^2 test is the equivalent of a two-sided test using the standard normal distribution. In our example, we note that $Z^2 = (.376)^2 = .141$, which is equal to $\chi^2 = .141$.

6.10 Test for Homogeneity of Several Values of r

As we have stated previously, the χ^2 test is applicable to the case of $k > 2$ independent random values of r. Table 6.2 gives the sample sizes and values of

TABLE 6.2   Calculation of the χ^2 Test of the Homogeneity of $k = 4$ Independent Values of *r*

	(1) n_i	(2) r_i	(3) $n_i - 3$	(4) z_i	(5) $(n_i - 3)z_i$	(6) $(n_i - 3)z_i^2$
	33	.53	30	.590	17.700	10.443
	58	.62	55	.725	39.875	28.909
	42	.65	39	.775	30.225	23.424
	47	.45	42	.485	21.340	10.350
Σ	180	2.25	168	2.575	109.140	73.126

r for each of $k = 4$ independent random samples. The z_i values for each value of *r* are given in column (4). Column (5) gives the values of $(n_i - 3)z_i$ and column (6) gives the values of $(n_i - 3)z_i^2$. In general, for *k* independent values of *r*,

$$\chi^2 = \sum_1^k (n_i - 3)z_i^2 - \frac{\left[\sum_1^k (n_i - 3)z_i\right]^2}{\sum_1^k (n_i - 3)} \tag{6.9}$$

with $k - 1$ d.f.

Substituting the appropriate values from Table 6.2 in (6.9), we have

$$\chi^2 = 73.126 - \frac{(109.140)^2}{168} = 2.22$$

a nonsignificant value with $k - 1 = 3$ d.f. and with $\alpha = .05$. We thus conclude that the various *r*'s can be considered estimates of the same population value ρ. Therefore, we calculate

$$\bar{z}_r = \frac{\sum_1^k (n_i - 3)z_i}{\sum_1^k (n_i - 3)} \tag{6.10}$$

or the weighted average value of z_r. For the data of Table 6.2, we have

$$\bar{z}_r = \frac{109.140}{168} = .650$$

From Table V in the Appendix we find that the corresponding value of *r* is approximately .570. We may regard this value as an estimate of the common population value ρ based on the combined data from the four samples.

6.11 Test of Significance of a Partial Correlation Coefficient: $r_{12.3}$

If we have three variables and we know the values of their intercorrelations, then we have defined the partial correlation coefficient between X_1 and X_2 with X_3 held constant as

$$r_{12.3} = \frac{r_{12} - r_{13}r_{23}}{\sqrt{1 - r_{13}^2}\sqrt{1 - r_{23}^2}}$$

To test the null hypothesis that $\rho_{12.3} = 0$, we may either calculate

$$t = \frac{r_{12.3}}{\sqrt{1 - r_{12.3}^2}}\sqrt{n - 3} \tag{6.11}$$

with $n - 3$ d.f. or use Table IV in the Appendix with $n - 3$ d.f. rather than $n - 2$ d.f.

Exercises

6.1 For a random sample of $n = 10$, we have $r = .88$. Test the null hypothesis that $\rho = 0$. Make a two-sided test with $\alpha = .05$.

6.2 With a sample of $n = 20$, what value of r would be required in order to reject the null hypothesis that $\rho = 0$, if a two-sided test is made with $\alpha = .05$?

6.3 For a sample of $n = 25$, we have $r = .82$. Find a 95 percent confidence interval for ρ.

6.4 The correlation coefficient between scores on an algebra test and grades on a final examination for one section of an elementary statistics class consisting of 48 students was .56. For another section of 44 students the correlation coefficient was .45. Do these two values of r differ significantly? Make the test using both Z and χ^2. You should find that $Z^2 = \chi^2$.

6.5 For three independent random samples the correlation coefficients between scores on two personality scales were .83, .70, and .90. The number of observations on which the correlation coefficients were based were 28, 39, and 52, respectively. (a) Determine whether the correlation coefficients differ significantly, with $\alpha = .05$. (b) If they do not differ significantly, what is the estimated value of ρ based on the combined samples?

6.6 Briefly define each of the following concepts or terms:

confidence interval
Type I error
significance level of a test
null hypothesis
bivariate normal distribution

Seven

Tests of Significance for Special Cases of the Correlation Coefficient

7.1 Introduction

When X and Y have, in the population, a bivariate normal distribution, then we have seen that it is possible to make various tests of significance regarding the population value ρ. For example, we could test a null hypothesis $\rho = .50$ for a single sample or a null hypothesis $\rho_1 = \rho_2$ for two independent random samples, or that $\rho_1 = \rho_2 = \cdots = \rho_k$ for k independent random samples.

In the special cases of the correlation coefficient, described in Chapter 5, the joint distribution of X and Y cannot be a bivariate normal distribution. This is obviously impossible in the case of the correlation coefficient between two dichotomous variables, and in the case of a dichotomous variable and another variable that can take any number of different values. It is also impossible in the case of the correlation coefficient between two sets of ranks. This means that, in general, the various tests of significance we have described with respect to correlation coefficients for bivariate normally distributed variables do not apply in the special cases of the correlation coefficient described in Chapter 5.

We can, however, show that if the Y values are independent of the X values, then the phi coefficient, the point biserial coefficient, and the rank order correlation coefficient must, of necessity, be equal to zero. Thus any positive or negative value of r for any of these three special cases may be assumed to offer evidence that the Y values are not independent of the X values. If the obtained value of r in these special cases is sufficiently large that we reject the null hypothesis of independence, then we may conclude that the Y values are dependent on the X values or, in other words, that there is a significant association between the X and Y values.

7.2 Test of Significance for the Phi Coefficient

We consider first the correlation coefficient between two dichotomous variables. It will simplify the discussion to consider two items, X and Y, the response to each of which may be either correct or incorrect. We assign a value of 1 to a correct response and a value of 0 to an incorrect response. Figure 7.1(a) shows, for a sample of $n = 200$ individuals, $p_1 = .50$, the proportion giving the correct response to Item 1 (X), and $p_2 = .60$, the proportion giving the correct response to Item 2 (Y). We note that $q_1 = 1 - p_1$ and that $q_2 = 1 - p_2$. The cells of the figure give the possible response patterns $(1,0)$, $(1,1)$, $(0,0)$, and $(0,1)$. If the sample of subjects is random, then p_1, which is the mean value of X, can be shown to be an unbiased estimate of the population mean P_1, and P_1 is simply the probability that $X = 1$ if an observation is drawn at random from the population of X values. Similarly, p_2 is an unbiased estimate of P_2 and P_2 is simply the probability that $Y = 1$ if an observation is drawn at random from the population of Y values.

We recall from elementary algebra that if we have two independent

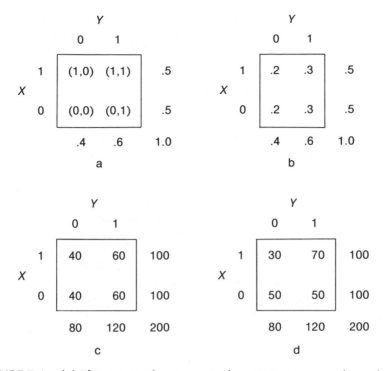

FIGURE 7.1   (a) The proportion p_1 = .5 of correct responses (X = 1) and the proportion p_2 = .6 of correct responses (Y = 1) to two items. (b) The probability of each response pattern shown in (a) occurring if responses to the two items are independent. (c) The expected frequency of each response pattern in a sample of n = 200 if responses to the two items are independent. (d) The observed frequency of each response pattern in a sample of n = 200.

events A and B and if $P(A)$ is the probability of A occurring and $P(B)$ is the probability of B occurring, then the probability that both A and B will occur will be given by the product of the two probabilities, that is, $P(A)P(B)$. Similarly, if response to Item 1 is independent of response to Item 2, then the probability that X = 1 *and* Y = 1 or that the response pattern (1,1) will occur will be given by $P_1 \times P_2$. We have, as sample estimates of P_1 and P_2, p_1 = .50 and p_2 = .60. Thus, if response to Item 1 is independent of response to Item 2, we have $p_1 p_2$ = (.5)(.6) = .30 as an estimate of the probability of the response pattern (1,1), $p_1 q_2$ = (.5)(.4) = .20 as an estimate of the probability of the response pattern (1,0), $q_1 p_2$ = (.5)(.6) = .30 as an estimate of the probability of the response pattern (0,1), and $q_1 q_2$ = (.5)(.4) = .20 as an estimate of the probability of the response pattern (0,0). These probabilities have been entered in the cells of Figure 7.1(b).

If we multiply the probabilities just obtained by the sample size, n = 200, we obtain the *expected* number of individuals who will have each of the response patterns; these frequencies or expected numbers are shown in Figure 7.1(*c*). If we now calculate the correlation coefficient for the entries in Figure 7.1(*c*), we obtain

$$r = \frac{(60)(40) - (40)(60)}{\sqrt{(100)(100)(80)(120)}} = 0$$

Thus, if *r* is equal to zero, the observed data are in accord with the notion that responses to the two items are independent. On the other hand, any positive or negative value of *r* will indicate that the responses to the two items are not independent, but are instead associated.

Now, even when the responses to the two items in the population are independent, the observed cell entries corresponding to the possible patterns of response will show some deviation from the expected frequencies, simply as a result of random sampling. We are interested in determining whether the deviations are sufficiently large to reject the hypothesis that, in the populations, responses to the two items are independent.

Assume, for example, that the observed frequencies are those shown in Figure 7.1(*d*) for a sample of $n = 200$. According to the hypothesis of independence, these observed frequencies, f_i, should deviate from the theoretical frequencies, F_i, only as a result of random sampling. As a test of the null hypothesis of independence, we calculate

$$\chi^2 = \sum \frac{(f_i - F_i)^2}{F_i} \tag{7.1}$$

and if the null hypothesis is true, χ^2 as obtained from (7.1) will be distributed as χ^2 with 1 d.f. If the obtained value of χ^2 exceeds the tabled value at, say, the $\alpha = .05$ significance level, the null hypothesis will be rejected. For the data in Figure 7.1(d), we have

$$\chi^2 = \frac{(30 - 40)^2}{40} + \frac{(70 - 60)^2}{60} + \frac{(50 - 40)^2}{40} + \frac{(50 - 60)^2}{60}$$
$$= 2.50 + 1.67 + 2.50 + 1.67$$
$$= 8.34$$

with 1 d.f. From the table of χ^2 we find that with 1 d.f., $P(\chi^2 \geq 3.841) = .05$, and we may conclude that the disagreement between the theoretical frequencies and the observed frequencies is such that the responses to the two items are not independent.

The χ^2 given by (7.1) is very simply related to the square of the correlation coefficient for two dichotomous variables of the kind described. In this particular case, it can be shown that

$$\chi^2 = nr^2 \tag{7.2}$$

In our example, we have

$$r^2 = \frac{[(70)(50) - (30)(50)]^2}{(80)(120)(100)(100)} = \frac{1}{24} = .0417$$

and, therefore,

$$\chi^2 = (200)(.0417) = 8.34$$

which is the same value we obtained using (7.1).

7.3 The *t* Test of Significance for the Point Biserial Coefficient

Table 7.1 gives the values of Y and Y^2, for a variable that can take a number of different values, in a sample in which $n_0 = 10$ observations have a value of $X = 0$ and $n_1 = 10$ observations have a value of $X = 1$. To determine the value of point biserial coefficient, we first find

$$SS_T = n_0(\overline{Y}_0 - \overline{Y})^2 + n_1(\overline{Y}_1 - \overline{Y})^2$$

$$= \frac{(\Sigma Y_0)^2}{n_0} + \frac{(\Sigma Y_1)^2}{n_1} - \frac{(\Sigma Y_0 + \Sigma Y_1)^2}{n_0 + n_1} \qquad (7.3)$$

or, for the data in Table 7.1,

$$SS_T = \frac{(30)^2}{10} + \frac{(80)^2}{10} - \frac{(110)^2}{20} = 125$$

TABLE 7.1   Values of Y and Y² for a Variable that Can Take a Number of Different Values in a Sample in Which $n_0 = 10$ Observations Have a Value of X = 0 and $n_1 = 10$ Observations Have a Value of X = 1

	Values of Y		Values of Y^2	
	X = 0	X = 1	X = 0	X = 1
	4	8	16	64
	3	10	9	100
	3	10	9	100
	1	9	1	81
	2	6	4	36
	5	7	25	49
	5	7	25	49
	4	9	16	81
	1	6	1	36
	2	8	4	64
Σ	30	80	110	660

We also have

$$SS_{tot} = 770 - \frac{(110)^2}{20} = 165$$

Then, for the value of the point biserial coefficient, we obtain

$$r = \sqrt{\frac{SS_T}{SS_{tot}}} = \sqrt{\frac{125}{165}} = \sqrt{.7576} = .8704$$

To test the null hypothesis that in the population ρ is equal to zero, we can calculate

$$t = \frac{r}{\sqrt{1 - r^2}} \sqrt{n - 2} \tag{7.4}$$

In our example we have

$$t = \frac{.8704}{\sqrt{1 - .7576}} \sqrt{20 - 2} = 7.50$$

with 18 d.f. With 18 d.f., we find from the table of t that $P(t \geq 7.50)$ or $P(t \leq -7.50)$ is less than .01, and the null hypothesis is thus rejected.

We could also use Table IV in the Appendix to determine whether $r = .8704$ is significant. For a two-sided test, with $\alpha = .01$, and with 18 d.f., we find that either $r \geq .561$ or $r \leq -.561$ would result in the rejection of the null hypothesis. Because our obtained value of the point biserial coefficient, .8704, exceeds the tabled value, the null hypothesis would be rejected.

In one of the formulas we developed for the point biserial coefficient, we showed that if $\overline{Y}_0$ is equal to $\overline{Y}_1$, then r must be equal to zero. If the sample is random, then $\overline{Y}_0$ is an unbiased estimate of the population mean μ_0 and $\overline{Y}_1$ is an unbiased estimate of the population mean μ_1. In the case of a point biserial coefficient, the t test of the null hypothesis that $\rho = 0$, as given by (7.4), is equivalent to a test of the null hypothesis that in the population $\mu_0 = \mu_1$. The only assumptions involved in the t test for the point biserial coefficient are that the Y values for each value of X are normally distributed about the corresponding population means and that the population variance of the Y values is the same for each of the X values.

Under the assumption that $\sigma_{Y.0}^2 = \sigma_{Y.1}^2 = \sigma_Y^2$, our best estimate of the common population variance will be given by

$$s_Y^2 = \frac{\Sigma y_0^2 + \Sigma y_1^2}{n_0 + n_1 - 2} \tag{7.5}$$

For the data in Table 7.1, we have

$$\Sigma y_0^2 = 110 - \frac{(30)^2}{10} = 20$$

and

$$\Sigma y_1^2 = 660 - \frac{(80)^2}{10} = 20$$

Then

$$s_Y^2 = \frac{20 + 20}{10 + 10 - 2} = 2.2222$$

If the null hypothesis $\mu_0 = \mu_1$ is true, then

$$t = \frac{\overline{Y}_1 - \overline{Y}_0}{\sqrt{s_Y^2\left(\frac{1}{n_0} + \frac{1}{n_1}\right)}} \tag{7.6}$$

will have a t distribution with $n_0 + n_1 - 2$ d.f. Substituting $\overline{Y}_1 = 8, \overline{Y}_0 = 3, s_Y^2 = 2.2222$, and $n_0 = n_1 = 10$ in (7.6), we obtain

$$t = \frac{8 - 3}{\sqrt{2.2222\left(\frac{1}{10} + \frac{1}{10}\right)}} = 7.50$$

which is equal to the value of t we obtained using (7.4). We conclude that $\mu_0 \neq \mu_1$ or, in other words, that the values of the Y means are not independent of the X classification.

7.4 The F Test of Significance for the Point Biserial Coefficient

A test of the null hypothesis $\mu_0 = \mu_1$ may also be made in terms of the F distribution. If the null hypothesis is true, then it can be shown that

$$F = \frac{SS_T}{s_Y^2} \tag{7.7}$$

where s_Y^2 is defined by (7.5), will be distributed as F with 1 d.f. for the numerator and $n_0 + n_1 - 2$ d.f. for the denominator.

In our example, we have $SS_T = 125$ and $s_Y^2 = 2.2222$. Then

$$F = \frac{125}{2.2222} = 56.25$$

with 1 and 18 d.f. From the table of F, Table VI in the Appendix, we find that for 1 and 18 d.f., $P(F \geq 56.25)$ is less than .01 and the null hypothesis would be rejected.

We note that the square of t, as defined by (7.6), will always be equal to F, as defined by (7.7). In our example, we have $t^2 = (7.50)^2 = 56.25$ and this is equal to the value of F.

Regardless of whether a t test or an F test is made, a significant value indicates that the values of the Y means, $\overline{Y}_0$ and $\overline{Y}_1$, are not independent of the X classification.

7.5 Test of Significance for the Rank Order Correlation Coefficient

The value of the rank order correlation coefficient can be regarded as indicating whether one set of ranks is independent of the other set of ranks or whether there is some agreement between the two sets of rankings. The distribution of the rank order correlation coefficient could, in theory, be calculated for any two sets of n ranks. We could, for example, arrange the ranks for one variable in standard order from 1 to n. We could then take every possible permutation of the ranks for the other variable and correlate each of the resulting permutations with the standard order for the first set of ranks.[1] According to the null hypothesis that the two sets of ranks are independent, each of these permutations is equally likely.

Olds[2] tabled the values of ΣD^2 for n ranks from 2 through 7 in terms of exact frequencies in the manner just described and for n equal to 8, 9, and 10 by means of an approximation function. We have used his table to calculate the corresponding values of r. These values are given in Table VIII in the Appendix, which enables us to test the null hypothesis that the ranks are independent for relatively small values of n. The first column of Table VIII gives the value of n, the number of ranks from 4 to 10, and the second column gives selected values of r for these sets of ranks. The third column gives the probability of obtaining r equal to or greater than the value given in column 2, when the null hypothesis is true. The tabled values of r are thus for a one-sided test. For example, with $n = 7$, the probability of $r \geq .679$ is equal to .0548, and this is also the probability of obtaining $r \leq -.679$, when the null hypothesis of independence is true. The probability of $r \geq .679$ or $r \leq -.679$ for a two-sided test will be equal to $(2)(.0548) = .1096$.

When n is greater than 10, the sampling distribution of the rank order coefficient, under the null hypothesis that $\rho = 0$, may be approximated by the t distribution. Thus

$$t = \frac{r}{\sqrt{1 - r^2}} \sqrt{n - 2}$$

with $n - 2$ d.f. provides a test of the null hypothesis that $\rho = 0$. It is, however, not necessary to calculate t to determine whether r is significant. Instead, we

---

[1]We, in essence, regard one set of ranks as fixed and the other as random.
[2]Olds, E. G. Distributions of sums of rank differences for small numbers of individuals. *Annals of Mathematical Statistics*, 1938, *9*, 133–148.

may use Table IV in the Appendix, with $n - 2$ d.f., and evaluate the rank order correlation coefficient in terms of the tabled values in the same manner in which this table was used to test the significance of the point biserial coefficient.

Exercises

7.1 In a test consisting of two items, the proportion answering True to Item 1 is .60 and the proportion answering True to Item 2 is .70. (a) If responses to the two items are independent, what is the estimated probability of each of the response patterns (1,0), (1,1), (0,0), and (0,1) occurring? (b) If the sample consists of $n = 50$ individuals, what are the expected frequencies of each of the response patterns? (c) If the observed frequency of the response pattern (1,1) is 28, will responses to the two items be judged to be independent?

7.2 The scores of eight males and eight females on a test (Y) are as follows:

Males	Females
7	7
5	7
6	10
3	6
4	5
4	4
1	8
2	9

(a) Use a t test to determine whether the Y means are independent of the sex classification. (b) Use an F test to determine whether the Y means are independent of the sex classification. Note that $F = t^2$.

7.3 If the rank order correlation coefficient for a sample of $n = 8$ is equal to .80, would the null hypothesis $\rho = 0$ be rejected for a two-sided test with $\alpha = .05$?

7.4 For a sample of $n = 200$ the phi coefficient between responses to Item 1 and Item 2 is .23. Are responses to the two items independent?

Eight

Tests of Significance for Regression Coefficients

8.1 Introduction

As we have pointed out previously, in an experiment the values of X, the independent variable, are usually preselected by the experimenter and may be regarded as fixed. The X values, in other words, are not subject to random variation. For each fixed value of X we have a random sample of one or more observations of a dependent variable Y. If we have calculated a regression coefficient b_Y for a set of paired (X,Y) values, where the X values are fixed, we may be interested in determining whether the obtained value of b_Y differs significantly from zero; that is, we may wish to test the null hypothesis that the population parameter β is equal to zero.[1]

In this chapter we show how the t distribution may be used to test various null hypotheses regarding β and also to test the null hypothesis that $\beta_1 - \beta_2 = 0$, if we have two independent samples of paired (X,Y) values. In addition, we shall see how to apply a test of significance to determine whether three or more sample regression coefficients based on independent random samples can be assumed to be estimates of a common population value. The only distribution assumptions of these tests of significance for regression coefficients are those involving the Y variable.

It is assumed that there is a normally distributed population of Y values with the same variance $\sigma^2_{Y.X}$ for each fixed value of X. The ordered values of $\mu_{Y.X}$, the population means of the Y values for each of the X values, are assumed to fall on a straight line with slope equal to β. Note that the assumptions of tests of significance of regression coefficients are considerably less restrictive than those of tests of significance of correlation coefficients. We do not, for example, have to assume that the joint distribution of X and Y is bivariate normal.

In regression problems in which the X values are fixed, the relationship between the Y and X values may take forms other than a linear relationship. Methods for studying other than linear relationships will be discussed in Chapter 14. In this chapter we are concerned solely with tests of significance of regression coefficients for linear relationships.

8.2 Test of the Null Hypothesis that $\beta = 0$

Table 8.1 gives the X and Y values for two independent random samples of $n_1 = 10$ and $n_2 = 10$ observations. For the first sample, we have

$$\Sigma(X - \overline{X})^2 = 385 - \frac{(55)^2}{10} = 82.5$$

$$\Sigma(Y - \overline{Y})^2 = 694 - \frac{(78)^2}{10} = 85.6$$

[1] In this chapter we shall be concerned only with the regression coefficient b_Y and the subscript will be omitted in the discussion that follows.

TABLE 8.1   Values of (X, Y) in Two Independent Samples of n = 10 Observations Each

	Sample 1				
	X	Y	X^2	Y^2	XY
	1	2	1	4	2
	2	4	4	16	8
	3	8	9	64	24
	4	6	16	36	24
	5	8	25	64	40
	6	10	36	100	60
	7	8	49	64	56
	8	9	64	81	72
	9	12	81	144	108
	10	11	100	121	110
Σ	55	78	385	694	504

	Sample 2				
	X	Y	X^2	Y^2	XY
	1	6	1	36	6
	2	4	4	16	8
	3	8	9	64	24
	4	6	16	36	24
	5	8	25	64	40
	6	10	36	100	60
	7	12	49	144	84
	8	10	64	100	80
	9	11	81	121	99
	10	13	100	169	130
Σ	55	88	385	850	555

and

$$\Sigma(X - \overline{X})(Y - \overline{Y}) = 504 - \frac{(55)(78)}{10} = 75.0$$

Then, for the value of the regression coefficient, we obtain

$$b = \frac{75.0}{82.5} = .91$$

and the residual sum of squares will be given by

$$\Sigma(Y - Y')^2 = 85.6 - \frac{(75.0)^2}{82.5} = 17.42$$

The residual variance is equal to

$$s_{Y.X}^2 = \frac{17.42}{10 - 2} = 2.18$$

The standard error of the regression coefficient will then be given by

$$s_b = \frac{s_{Y.X}}{\sqrt{\Sigma x^2}} \tag{8.1}$$

or

$$s_b = \sqrt{\frac{s_{Y.X}^2}{\Sigma x^2}} \tag{8.2}$$

and, in our example, we have

$$s_b = \sqrt{\frac{2.18}{82.5}} = \sqrt{.0264} = .1625$$

To test a null hypothesis regarding the value of β, we define

$$t = \frac{b - \beta}{s_b} \tag{8.3}$$

and if the null hypothesis is true, then the t defined by (8.3) will have a t distribution with $n - 2$ d.f. Specifically, if we wish to test the null hypothesis that $\beta = 0$, then, in our example, we have

$$t = \frac{.91}{.1625} = 5.60$$

with $10 - 2 = 8$ d.f. For a two-sided test of significance with 8 d.f. and with $\alpha = .05$, we find that $t \geq 2.306$ or $t \leq -2.306$ will be judged significant.

As we pointed out in our earlier discussion of the correlation coefficient, if b is equal to zero, then r must also be equal to zero. The t test of the null hypothesis that $\beta = 0$ is the same as the t test of the null hypothesis that $\rho = 0$. For the sample under discussion, we have

$$r = \frac{75.0}{\sqrt{(82.5)(85.6)}} = .8925$$

and, as a test of the null hypothesis that $\rho = 0$, we have

$$t = \frac{.8925}{\sqrt{1 - (.8925)^2}} \sqrt{10 - 2} = 5.60$$

which is the same value of t we obtained when we tested the null hypothesis that $\beta = 0$. The proof that these two values of t must be equal is given in the answer to one of the exercises at the end of this chapter.

We see, therefore, that the t test of the null hypothesis $\rho = 0$ requires no assumptions other than those of the t test of the null hypothesis $\beta = 0$. This is also true of the t test of the null hypothesis $\rho = 0$ for a point biserial coefficient.

For the second sample of $n_2 = 10$ observations, we have

$$\Sigma(X - \overline{X})^2 = 385 - \frac{(55)^2}{10} = 82.5$$

$$\Sigma(Y - \overline{Y})^2 = 850 - \frac{(88)^2}{10} = 75.6$$

and

$$\Sigma(X - \overline{X})(Y - \overline{Y}) = 555 - \frac{(55)(88)}{10} = 71.0$$

Then, the regression coefficient for the second sample is

$$b = \frac{71.0}{82.5} = .86$$

and for the sum of squared errors of prediction, we have

$$\Sigma(Y - Y')^2 = 75.6 - \frac{(71.0)^2}{82.5} = 14.5$$

For the residual variance for this sample, we have

$$s_{Y.X}^2 = \frac{14.5}{10 - 2} = 1.81$$

and for the standard error of b, we obtain

$$s_b = \sqrt{\frac{1.81}{82.5}} = \sqrt{.0219} = .1480$$

and

$$t = \frac{.86}{.1480} = 5.81$$

which is also a significant value with $\alpha = .05$ and with $10 - 2 = 8$ d.f.

8.3 Test of the Null Hypothesis that $\beta_1 - \beta_2 = 0$

We now wish to determine whether $b_1 = .91$ and $b_2 = .86$ differ significantly, that is, we wish to test the null hypothesis that $\beta_1 - \beta_2 = 0$. We assume homogeneity of the residual variances for the two groups on the Y variable and obtain as our estimate of the common residual variance[2]

$$s_{Y.X}^2 = \frac{\left[\Sigma y_1^2 - \frac{(\Sigma x y_1)^2}{\Sigma x_1^2}\right] + \left[\Sigma y_2^2 - \frac{(\Sigma x y_2)^2}{\Sigma x_2^2}\right]}{n_1 + n_2 - 4} \tag{8.4}$$

Substituting the values we have already calculated in (8.4), we have

$$s_{Y.X}^2 = \frac{\left[85.6 - \frac{(75.0)^2}{82.5}\right] + \left[75.6 - \frac{(71.0)^2}{82.5}\right]}{10 + 10 - 4}$$

$$= \frac{17.42 + 14.50}{16}$$

$$= 1.995$$

The standard error of the difference between the two regression coefficients will be given by

$$s_{b_1 - b_2} = \sqrt{s_{Y.X}^2 \left(\frac{1}{\Sigma x_1^2} + \frac{1}{\Sigma x_2^2}\right)} \tag{8.5}$$

where $s_{Y.X}^2$ is defined by (8.4) and, in our example, is equal to 1.995. Substituting $\Sigma x_1^2 = \Sigma x_2^2 = 82.5$ in (8.5), we have

$$s_{b_1 - b_2} = \sqrt{1.995 \left(\frac{1}{82.5} + \frac{1}{82.5}\right)} = .22$$

We then define

$$t = \frac{(b_1 - b_2) - (\beta_1 - \beta_2)}{s_{b_1 - b_2}} \tag{8.6}$$

The t defined by (8.6) will have a t distribution with $n_1 + n_2 - 4$ d.f., when the null hypothesis is true, and can be evaluated in terms of the table of t. In our example, we have $b_1 = .91$ and $b_2 = .86$ with $s_{b_1 - b_2} = .22$. As a test of the null hypothesis that $\beta_1 - \beta_2 = 0$, we have

$$t = \frac{.91 - .86}{.22} = .23$$

From the table of t we find that for 16 d.f. and with a two-sided test, $t \geq 2.12$ or $t \leq -2.12$ will be significant with $\alpha = .05$, when the null hypothesis is true. Because our obtained value of t is equal to .23, we can regard the null hypothesis $\beta_1 - \beta_2 = 0$ as tenable.

---

[2]We have $s_{Y.X}^2 = 2.18$ for the first sample and $s_{Y.X}^2 = 1.81$ for the second sample. Then, as a test of the validity of the assumption that the two variance estimates are homogeneous, we have $F = 2.18/1.81 = 1.20$ with 8 and 8 d.f., and this is a nonsignificant value of F with $\alpha = .01$.

The test of the null hypothesis that $\beta_1 - \beta_2 = 0$ is *not* equivalent to the test of the null hypothesis that $\rho_1 - \rho_2 = 0$. The test of the null hypothesis $\beta_1 - \beta_2 = 0$ is a test to determine whether the slopes of the two regression lines differ significantly. The slopes may differ significantly, whereas the two correlation coefficients may not. Similarly, the slopes of the two regression lines may not differ significantly, whereas the two correlation coefficients may differ significantly.

8.4 Test for Homogeneity of Several Independent Values of *b*

In some experiments, we may have three or more independent random samples. For each of the samples we may calculate the regression coefficient *b*. We are interested in determining whether the several sample values of *b* may be regarded as estimating the same common population value β. A test of the null hypothesis that $\beta_1 = \beta_2 = \cdots = \beta_k$ can be made in terms of the F distribution.

Table 8.2 gives the values of Σx^2, Σxy, Σy^2, b, $(\Sigma xy)^2/\Sigma x^2$, and

$$\Sigma(Y - Y')^2 = \Sigma y^2 - \frac{(\Sigma xy)^2}{\Sigma x^2}$$

for each of three independent samples of $n = 5$ observations each.[3]

The residual sums of squares $\Sigma(Y - Y')^2$ shown in column (7) of the table for each of the three samples have been minimized because for each sample we used the regression coefficient unique to the sample in obtaining the residual sum of squares. The sum of the three residual sums of squares is equal to 9.66, and this sum of squares is designated as SS_1 in the table. For SS_1 we have $k(n - 2)$ d.f., where k is the number of samples and n is the

---

[3] The test of significance does not require that we have the same number of observations in each group.

TABLE 8.2   Values of Σx^2, Σxy, Σy^2, b, $(\Sigma xy)^2/\Sigma x^2$, and $\Sigma(Y - Y')^2$ for Three Independent Samples of $n = 5$ Observations Each

(1) Sample	(2) Σx^2	(3) Σxy	(4) Σy^2	(5) b	(6) $(\Sigma xy)^2/\Sigma x^2$	(7) $\Sigma(Y - Y')^2$	(8) d.f.
(1) Sample 1	15.00	20.00	30.00	1.33	26.67	3.33	3
(2) Sample 2	15.00	23.00	38.00	1.53	35.27	2.73	3
(3) Sample 3	15.00	21.00	33.00	1.40	29.40	3.60	3
						$SS_1 = 9.66$	9
(4) Pooled	45.00	64.00	101.00	1.42	91.02	$SS_2 = 9.98$	11
						$SS_3 = SS_2 - SS_1 = 0.32$	2

number of observations in each sample. Thus, in this problem, we have $3(5 - 2) = 9$ d.f. for SS_1.

Let us assume, for the moment, that the separate regression coefficients are all estimates of the same common population regression coefficient. Then an estimate of this common regression coefficient, which we designate as b_w, will be

$$b_w = \frac{\Sigma xy_1 + \Sigma xy_2 + \Sigma xy_3}{\Sigma x_1^2 + \Sigma x_2^2 + \Sigma x_3^2} \tag{8.7}$$

or, in our example,

$$b_w = \frac{20.00 + 23.00 + 21.00}{15.0 + 15.0 + 15.0} = \frac{64.0}{45.0} = 1.42$$

Then we can also obtain a residual sum of squares representing the squared deviations $(Y - Y')^2$ when a line with a common slope equal to $b_w = 1.42$ is used for all three samples. We designate this residual sum of squares as SS_2 and, in our example,

$$SS_2 = 101.0 - \frac{(64.0)^2}{45.0} = 9.98$$

as shown in row (4) and column (7) of Table 8.2. SS_2 will have $k(n - 1) - 1$ d.f., or, in our example, $3(5 - 1) - 1 = 11$ d.f.[4]

Now SS_2 can never be smaller than SS_1, because SS_1 is based on the squared deviations within each sample from a regression line with slope b_i fitted separately for each sample, and the values of b_i were found in such a way as to minimize the sum of squared deviations within each sample. SS_2, on the other hand, is based on the squared deviations from a regression line that has the same slope, b_w, for each sample. Thus if the regression coefficient b_i for a given sample is not equal to b_w, the sum of squared deviations for this group from the line with slope b_w will be larger than that from the line with slope b_i. If $b_1 = b_2 = b_3 = b_w$, then SS_2 will be exactly equal to SS_1. If the b_i values do show considerable variation, then SS_2 will be considerably larger than SS_1.

To determine whether the regression coefficients differ significantly, we find

$$SS_3 = SS_2 - SS_1 \tag{8.8}$$

SS_3 will have $k - 1$ d.f., where k is the number of samples or separate regression coefficients. In our example, we have

$$SS_3 = 9.98 - 9.66 = .32$$

with $k - 1 = 2$ d.f.

---

[4]If the number of observations is not the same for the various samples, then the degrees of freedom for SS_1 will be given by $\Sigma_1^k n_i - 2k$, where n_i is the number of observations in a sample and k is the number of samples. For SS_2 the degrees of freedom will be given by $\Sigma_1^k n_i - k - 1$.

For a test of significance of the differences among the regression coefficients, we have

$$F = \frac{SS_3/(k-1)}{SS_1/k(n-2)} \qquad (8.9)$$

or, for the present problem,

$$F = \frac{.32/2}{9.66/9} = \frac{.16}{1.07} = .15$$

with 2 and 9 d.f., and this is a nonsignificant value. We conclude that the data offer no significant evidence against the null hypothesis $\beta_1 = \beta_2 = \beta_3$. Our sample estimate of β is $b_w = 1.42$.

Exercises

8.1   We have a sample of $n = 5$ paired (X,Y) values as follows:

X	Y
5	10
4	6
3	4
2	4
1	1

(a) Calculate the value of the regression coefficient of Y on X. (b) Calculate $s_{Y.X}^2$ and the standard error s_b. (c) Test the null hypothesis that $\beta = 0$, using a two-sided t test with $\alpha = .05$.

8.2   We have another independent random sample of $n = 5$ paired (X,Y) values as follows:

X	Y
5	3
4	3
3	1
2	2
1	1

(a) Calculate the value of the regression coefficient of Y on X. (b) Calculate $s_{Y.X}^2$ and s_b. (c) Test the null hypothesis that $\beta = 0$ using a two-sided t test with $\alpha = .05$.

8.3   Use the data in Exercises 8.1 and 8.2 and assume that $s_{Y.X}^2$ for each sample is an estimate of the same parameter $\sigma_{Y.X}^2$. (a) Pool the residual sums of squares and find $s_{Y.X}^2$ for the combined samples. (b) Find the standard error of the difference between the two regression coefficients. (c) Test the null hypothesis that $\beta_1 - \beta_2 = 0$.

8.4   Use the two samples given in Exercises 8.1 and 8.2. (a) Calculate SS_1, SS_2, and SS_3. (b) Use the F test to determine whether b_1 and b_2 differ significantly. (c) Show that the value of F obtained is equal to the square of the value of t obtained in Exercise 8.3.

8.5   If one regression coefficient, b_1, based on one sample differs significantly from zero and another regression coefficient, b_2, based on an independent sample does not differ significantly from zero, does it necessarily follow that b_1 and b_2 will differ significantly? Explain your answer.

8.6   If both b_1 and b_2 for two independent samples differ significantly from zero, does it necessarily follow that the difference between b_1 and b_2 will not be significant? Explain your answer.

8.7   If b is equal to zero, will r also be equal to zero? Explain why or why not.

8.8   Prove that the t test of the null hypothesis that $\rho = 0$, that is,

$$t = \frac{r}{\sqrt{1 - r^2}} \sqrt{n - 2}$$

is algebraically equivalent to the t test of the null hypothesis that $\beta = 0$, that is,

$$t = \frac{b}{\sqrt{s_{Y.X}^2 / \Sigma x^2}}$$

8.9   If we reject the null hypothesis $\beta_1 - \beta_2 = 0$ for two independent samples, would we also reject the null hypothesis $\rho_1 - \rho_2 = 0$, at the same significance level and for the same samples? Explain your answer.

8.10.   What information is provided by the test of the null hypothesis $\beta_1 - \beta_2 = 0$ that is not provided by the test of the null hypothesis $\rho_1 - \rho_2 = 0$?

8.11   What information is provided by the test of the null hypothesis $\rho_1 - \rho_2 = 0$ that is not provided by the test of the null hypothesis $\beta_1 - \beta_2 = 0$?

8.12   If we have a typical correlation problem with a random sample of paired (X,Y) values, and if we reject the null hypothesis $\rho = 0$, would we also reject the null hypotheses $\beta_y = 0$ and $\beta_x = 0$? Explain why or why not.

Nine

Multiple Correlation and Regression

9.1 Introduction

Suppose that for a random sample of n subjects we have available for each subject measures of three or more variables. One of the variables is of primary interest in that we would like to predict it from some weighted linear combination of the other variables. We shall refer to the variable we are interested in predicting as the Y variable and the remaining k variables as the X variables. Thus, we have a multiple regression equation

$$Y' = a + b_1X_1 + b_2X_2 + \cdots + b_kX_k$$

and we want to find the values of $a, b_1, b_2, \ldots, b_k$ that will result in the highest possible positive correlation between the observed values of Y and the predicted values or Y'. If this is done, then the resulting correlation coefficient between Y and Y' is called a *multiple correlation coefficient*[1] and is represented by $R_{YY'}$.

As the number of X variables increases, the calculations involved in finding the values of $b_1, b_2, \ldots, b_k$, become complex and overwhelming,[2] although they can be accomplished quickly and easily by a high-speed electronic computer. Fortunately, many of the principles of multiple regression and correlation can be illustrated with an example consisting of one Y variable and two X variables. The calculations used to solve the three-variable problem are relatively simple.

9.2 Calculating the Values of b_1 and b_2 for a Three-Variable Problem

With two X variables, the multiple regression equation becomes

$$Y' = a + b_1X_1 + b_2X_2 \tag{9.1}$$

and

$$Y = a + b_1X_1 + b_2X_2 + e \tag{9.2}$$

The maximum value of the correlation coefficient between the observed Y values and the predicted Y' values will be obtained if we find the values of a, b_1, and b_2 that minimize the sum of squared errors of prediction or the residual sum of squares

$$SS_{res} = \Sigma(Y - Y')^2 = \Sigma e^2$$

The required value of a is given by the equation

$$a = \overline{Y} - b_1\overline{X}_1 - b_2\overline{X}_2 \tag{9.3}$$

---

[1]The notation $R_{Y.123\ldots k}$ is also used for the multiple correlation coefficient.

[2]This statement is correct when the X variables are intercorrelated, as they typically are. The calculations are relatively simple when the X variables are mutually orthogonal, as we shall see later.

Substituting this value in (9.1), we obtain

$$Y' = \overline{Y} + b_1(X_1 - \overline{X}_1) + b_2(X_2 - \overline{X}_2)$$

or

$$Y' = \overline{Y} + b_1 x_1 + b_2 x_2$$

Then, if the residual sum of squares is to be minimized, the values of b_1 and b_2 must satisfy the following equations:

$$b_1 \Sigma x_1^2 + b_2 \Sigma x_1 x_2 = \Sigma x_1 y \tag{9.4}$$

and

$$b_1 \Sigma x_1 x_2 + b_2 \Sigma x_2^2 = \Sigma x_2 y \tag{9.5}$$

We have two equations with two unknowns, b_1 and b_2, and these equations can be solved by using standard algebraic methods. For example, if we multiply (9.4) by Σx_2^2 and (9.5) by $\Sigma x_1 x_2$, we obtain

$$b_1 (\Sigma x_1^2)(\Sigma x_2^2) + b_2 (\Sigma x_1 x_2)(\Sigma x_2^2) = (\Sigma x_1 y)(\Sigma x_2^2) \tag{9.6}$$

and

$$b_1 (\Sigma x_1 x_2)^2 + b_2 (\Sigma x_1 x_2)(\Sigma x_2^2) = (\Sigma x_2 y)(\Sigma x_1 x_2) \tag{9.7}$$

Then, subtracting (9.7) from (9.6), we have

$$b_1 (\Sigma x_1^2)(\Sigma x_2^2) - b_1 (\Sigma x_1 x_2)^2 = (\Sigma x_1 y)(\Sigma x_2^2) - (\Sigma x_2 y)(\Sigma x_1 x_2)$$

or

$$b_1 = \frac{(\Sigma x_1 y)(\Sigma x_2^2) - (\Sigma x_2 y)(\Sigma x_1 x_2)}{(\Sigma x_1^2)(\Sigma x_2^2) - (\Sigma x_1 x_2)^2} \tag{9.8}$$

Following a similar procedure, we find that

$$b_2 = \frac{(\Sigma x_2 y)(\Sigma x_1^2) - (\Sigma x_1 y)(\Sigma x_1 x_2)}{(\Sigma x_1^2)(\Sigma x_2^2) - (\Sigma x_1 x_2)^2} \tag{9.9}$$

9.3 A Numerical Example of a Three-Variable Problem

Assume that we have a random sample of $n = 20$ with one Y variable and two X variables.[3] Table 9.1 gives the values of

$$\Sigma x_1^2 = \Sigma(X_1 - \overline{X}_1)^2 = 76.0$$
$$\Sigma x_2^2 = \Sigma(X_2 - \overline{X}_2)^2 = 304.0$$
$$\Sigma y^2 = \Sigma(Y - \overline{Y})^2 = 1216.0$$

[3]In general, the number of observations n in a multiple correlation or regression problem should be considerably larger than the number of X variables k. No hard and fast rules are available, but we would suggest that n/k should be equal to or greater than 10.

TABLE 9.1   Sums of Squared Deviations and Sums of Products of Paired Deviations for Three Variables, X_1, X_2, and Y, for a Sample of $n = 20$ Observations

Variable	X_1	X_2	Y	Means
X_1	76.0	45.6	212.8	4.0
X_2		304.0	364.8	8.0
Y			1216.0	20.0
s	2.0	4.0	8.0	

$$\Sigma x_1 x_2 = \Sigma(X_1 - \overline{X}_1)(X_2 - \overline{X}_2) = 45.6$$
$$\Sigma x_1 y = \Sigma(X_1 - \overline{X}_1)(Y - \overline{Y}) = 212.8$$
$$\Sigma x_2 y = \Sigma(X_2 - \overline{X}_2)(Y - \overline{Y}) = 364.8$$

for the sample of $n = 20$ observations. The means of $\overline{X}_1$, $\overline{X}_2$, and $\overline{Y}$ are shown at the right of the table and the standard deviations appear at the bottom of the table. The diagonal entries in the table are the sums of squared deviations from the means. For example, we have $\Sigma x_1^2 = 76.0$. The nondiagonal entries give the sums of the products of the paired deviations. For example, we have $\Sigma x_1 x_2 = 45.6$. Substituting the appropriate values from Table 9.1 in (9.8) and (9.9), we have

$$b_1 = \frac{(212.8)(304.0) - (364.8)(45.6)}{(76.0)(304.0) - (45.6)^2} = 2.2857$$

and

$$b_2 = \frac{(364.8)(76.0) - (212.8)(45.6)}{(76.0)(304.0) - (45.6)^2} = .8571$$

Then, substituting in (14.3), we obtain

$$a = 20.0 - (2.2857)(4.0) - (.8571)(8.0) = 4.0$$

and the multiple regression equation will be

$$Y' = 4.0 + 2.2857X_1 + .8571X_2$$

9.4 Partitioning the Total Sum of Squares into the Regression and Residual Sums of Squares

We have previously shown that when there are two variables, X and Y, the total sum of squares for the Y variable or

$$SS_{tot} = \Sigma y^2 = \Sigma(Y - \overline{Y})^2$$

can be partitioned into two parts, the sum of squares for linear regression or

$$SS_{reg} = b\Sigma xy = \frac{(\Sigma xy)^2}{\Sigma x^2} = \Sigma(Y' - \bar{Y})^2$$

and the residual sum of squares

$$SS_{res} = \Sigma y^2 - \frac{(\Sigma xy)^2}{\Sigma x^2} = \Sigma(Y - Y')^2$$

This can also be done with a multiple regression equation, that is,

$$SS_{tot} = SS_{reg} + SS_{res} \tag{9.10}$$

In a multiple regression problem, the sum of squares for linear regression will be given by

$$SS_{reg} = b_1\Sigma x_1 y + b_2\Sigma x_2 y + \cdots + b_k\Sigma x_k y \tag{9.11}$$

In our example, we have two values of b with $b_1 = 2.2857$ and $b_2 = .8571$. Substituting these two values, along with $\Sigma x_1 y = 212.8$ and $\Sigma x_2 y = 364.8$, as given in Table 9.1, in (9.11), we have

$$SS_{reg} = (2.2857)(212.8) + (.8571)(364.8) = 799.07$$

as the sum of squares for linear regression with $k = 2$ d.f. Then the residual sum of squares can be obtained by subtraction, that is,

$$SS_{res} = SS_{tot} - SS_{reg} \tag{9.12}$$

and, in our example, we have

$$SS_{res} = 1216.00 - 799.07 = 416.93$$

with $n - k - 1 = 17$ d.f.

9.5 The Multiple Correlation Coefficient

The square of the multiple correlation coefficient will then be given by

$$R_{YY'}^2 = \frac{SS_{reg}}{SS_{tot}} \tag{9.13}$$

and, in our example, we have

$$R_{YY'}^2 = \frac{799.07}{1216.00} = .6571$$

The square root of $R_{YY'}^2$ is, of course, the multiple correlation coefficient, and

$$R_{YY'} = \sqrt{.6571} = .8106$$

Note also that

$$1 - R^2_{YY'} = 1 - \frac{SS_{reg}}{SS_{tot}} = \frac{SS_{tot} - SS_{reg}}{SS_{tot}} = \frac{SS_{res}}{SS_{tot}} \tag{9.14}$$

$R^2_{YY'}$ is, in other words, the proportion of SS_{tot} that can be accounted for by linear regression, and $1 - R^2_{YY'}$ is the proportion of SS_{tot} that is independent of the linear regression of Y on Y'.

9.6 Tests of Significance for R^2_{YY}

As (9.11) shows, if both b_1 and b_2 are equal to zero in a three-variable problem, then SS_{reg} will be equal to zero and $R^2_{YY'}$ will also be equal to zero. The obtained values of b_1 and b_2 are estimates of the corresponding population values β_1 and β_2. A test of the null hypothesis that the two population values β_1 and β_2 are *both* equal to zero will be given by

$$F = \frac{R^2_{YY'}/k}{(1 - R^2_{YY'})/(n - k - 1)} \tag{9.15}$$

or equivalently by

$$F = \frac{SS_{reg}/k}{SS_{res}/(n - k - 1)} = \frac{MS_{reg}}{MS_{res}} \tag{9.16}$$

For both F ratios the degrees of freedom for the numerator will be equal to k, the number of X variables, and the degrees of freedom for the denominator will be equal to $n - k - 1$.

Using (9.15), we have

$$F = \frac{.6571/2}{(1 - .6571)/17} = 16.29$$

or, equivalently, using (9.16), we have

$$F = \frac{799.07/2}{416.93/17} = 16.29$$

with 2 and 17 d.f. This is a significant value with $\alpha = .05$.

Unfortunately, the test of the null hypothesis $\beta_1 = \beta_2 = 0$ is not very satisfying. If the null hypothesis is rejected, then it may be because either β_1 or β_2 is not equal to zero or because both β_1 and β_2 are not equal to zero. There is a way, however, to test the null hypothesis $\beta_1 = 0$, given that X_2 has already been included in the regression equation, and also the null hypothesis $\beta_2 = 0$, given that X_1 has already been included in the regression equation. These conditional tests of significance are described in the next section.

9.7 Conditional Tests of Significance

When both X_1 and X_2 are included in the regression equation, the sum of squares for linear regression is equal to $SS_{reg} = 799.07$. Now suppose we find the regression sum of squares when only X_1 is included in the regression equation. Using the appropriate values given in Table 9.1, we have

$$SS_{reg} = \frac{(\Sigma x_1 y)^2}{\Sigma x_1^2} = \frac{(212.8)^2}{76.0} = 595.84$$

with 1 d.f. The difference, $799.07 - 595.84 = 203.23$, will be a measure of the contribution of X_2 to the regression sum of squares, given that X_1 is already present in the regression equation. Then the null hypothesis $\beta_2 = 0$ can be tested by

$$F = \frac{203.23}{24.53} = 8.28$$

with 1 and 17 d.f. From the table of F in the Appendix we find that this is a significant value with $\alpha = .05$.

Similarly, to determine whether X_1 contributes significantly to the regression sum of squares, given that X_2 has already been included in the regression equation, we find the regression sum of squares when only X_2 is included in the regression equation. Using the appropriate values given in Table 9.1, we have

$$SS_{reg} = \frac{(\Sigma x_2 y)^2}{\Sigma x_2^2} = \frac{(364.8)^2}{304.0} = 437.76$$

with 1 d.f. Then the difference, $799.07 - 437.76 = 361.31$, will measure the contribution of X_1 to the regression sum of squares, given that X_2 has already been included in the regression equation. In this instance, we have

$$F = \frac{361.31}{24.53} = 14.73$$

with 1 and 17 d.f. and this is also a significant value with $\alpha = .05$. These tests of significance are summarized in Table 9.2.

With only two X variables, the standard error of a regression coefficient will be given by

$$s_{b_i} = \sqrt{\frac{MS_{res}}{\Sigma x_i^2 (1 - r_{12}^2)}} \qquad (9.17)$$

where r_{12}^2 is the squared correlation coefficient between X_1 and X_2. As shown in Section 9.3, we have $\Sigma x_1 x_2 = 45.6$, $\Sigma x_1^2 = 76.0$, and $\Sigma x_2^2 = 304.0$. Then the

TABLE 9.2   Conditional Tests of Significance for X_1 and X_2

Source	d.f.	Sum of Squares	Mean Square	F
X_1 and X_2	2	799.07		
X_1 alone	1	595.84		
X_2 after X_1	1	203.23	203.23	8.28
X_1 and X_2	2	799.07		
X_2 alone	1	437.76		
X_1 after X_2	1	361.31	361.31	14.73
Residual	17	416.93	24.53	

squared correlation coefficient between X_1 and X_2 will be

$$r_{12}^2 = \frac{(\Sigma x_1 x_2)^2}{(\Sigma x_1^2)(\Sigma x_2^2)} = \frac{(45.6)^2}{(76.0)(304.0)} = .09$$

We also have $MS_{res} = SS_{res}/(n - k - 1) = 416.93/17 = 24.5253$. Then the standard error of b_1 will be

$$s_{b_1} = \sqrt{\frac{24.5253}{(76.0)(1 - .09)}} = .5955$$

We have previously found that $b_1 = 2.2857$. Then for the t test of b_1 we have

$$t = \frac{b_1}{s_{b_1}} = \frac{2.2587}{.5955} = 3.8383$$

and $t^2 = (3.8383)^2 = 14.73$. We observe that the square of the t test for b_1 is equivalent to the F test of the regression sum of squares for X_1 when X_1 is entered last (after X_2) in the regression equation.

Similarly, we have

$$s_{b_2} = \sqrt{\frac{24.5253}{(304.0)(1 - .09)}} = .2977$$

Then, for the t test of $b_2 = .8571$, we have

$$t = \frac{.8571}{.2977} = 2.8791$$

and $t^2 = (2.8791)^2 = 8.29$ is equal, within rounding errors, to the F test of the regression sum of squares for X_2 when X_2 is entered last (after X_1) in the regression equation.

9.8 Semipartial Correlations and Multiple Correlation

In the preceding section we found that the sum of squares for regression when both X_1 and X_2 were included in the regression equation was equal to 799.07. Using X_1 alone the regression sum of squares was 595.84 and the difference, $799.07 - 595.84 = 203.23$, was said to represent the unique contribution of X_2, given that X_1 was already included in the regression equation. If we divide 203.23 by $SS_{tot} = 1216.00$ and take the square root, we have

$$r_{Y(2.1)} = \sqrt{\frac{203.23}{1216.00}} = .4088$$

where $r_{Y(2.1)}$ is called a *semipartial* correlation coefficient. It is, in fact, the correlation between Y and X_2 after the variance that X_1 has in common with X_2 has been removed from X_2.

The semipartial correlation $r_{Y(2.1)}$ in our three-variable problem will also be given by[4]

$$r_{Y(2.1)} = \frac{r_{Y2} - r_{Y1}r_{12}}{\sqrt{1 - r_{12}^2}} \tag{9.18}$$

Substituting the appropriate values of the correlation coefficients in (9.18), we have

$$r_{Y(2.1)} = \frac{.60 - (.70)(.30)}{\sqrt{1 - (.30)^2}} = .4088$$

as before.

We know that $r_{Y1} = .70$ and that $(.70)^2 = .4900$ is the proportion of $\Sigma(Y - \overline{Y})^2$ that can be accounted for by the regression of Y on X_1. Given that we have taken X_1 into account, the additional proportion of $\Sigma(Y - \overline{Y})^2$ that can be accounted for by X_2 will be given by $r_{Y(2.1)}^2 = (.4088)^2 = .1671$. We now note that[5]

$$R_{YY'}^2 = R_{Y.12}^2 = r_{Y1}^2 + r_{Y(2.1)}^2$$

or, in our example,

$$R_{Y.12}^2 = .4900 + .1671 = .6571$$

---

[4]Note that the partial correlation coefficient

$$r_{Y2.1} = \frac{r_{Y2} - r_{Y1}r_{12}}{\sqrt{1 - r_{Y1}^2}\ \sqrt{1 - r_{12}^2}}$$

removes the variance of X_1 from both X_2 and Y. The semipartial correlation coefficient, on the other hand, removes the variance of X_1 only from X_2 and not from Y.

[5]With three X variables, it will also be true that

$$R_{Y.123}^2 = r_{Y1}^2 + r_{Y(2.1)}^2 + r_{Y(3.12)}^2$$

The semipartial correlation $r_{Y(3.12)}$ is the correlation between Y and X_3 after the variance that X_1 and X_2 have in common with X_3 has been removed from X_3.

The F test of $r^2_{Y(2.1)}$ will be given by

$$F = \frac{r^2_{Y(2.1)}}{(1 - R^2_{Y.12})/(n - k - 1)} \qquad (9.19)$$

In our example we have $r^2_{Y(2.1)} = (.4088)^2 = .1671$ and $R^2_{Y.12} = .6571$ with $n - k - 1 = 17$ d.f. Then

$$F = \frac{.1671}{(1 - .6571)/17} = 8.28$$

and we observe that the F test of $r^2_{Y(2.1)}$ is equivalent to the F test of the regression sum of squares for X_2 when it is entered last (after X_1) in the regression equation. The F test for $r^2_{Y(2.1)}$ is also equal to the square of the t test for b_2.

We found that the regression sum of squares, taking into account X_2 alone, was equal to 437.76. When both X_1 and X_2 were included in the regression equation, the regression sum of squares was equal to 799.07. Then the difference, $799.07 - 437.76 = 361.31$, was said to represent the unique contribution of X_1, given that X_2 had already been included in the regression equation. Thus we also have

$$r_{Y(1.2)} = \sqrt{\frac{361.31}{1216.00}} = .5451$$

The semipartial correlation coefficient in our three-variable problem will also be given by

$$r_{Y(1.2)} = \frac{r_{Y1} - r_{Y2}r_{12}}{\sqrt{1 - r^2_{12}}} \qquad (9.20)$$

Substituting the appropriate correlation coefficients in (9.20), we have

$$r_{Y(1.2)} = \frac{.70 - (.60)(.30)}{\sqrt{1 - (.30)^2}} = .5451$$

as before. Then it will also be true that

$$R^2_{YY'} = R^2_{Y.21} = r^2_{Y2} + r^2_{Y(1.2)}$$

We have $r_{Y2} = .60$ and $r_{Y(1.2)} = .5451$. Consequently,

$$R^2_{Y.21} = (.6000)^2 + (.5451)^2$$
$$= .3600 + .2971$$
$$= .6571$$

For the F test of significance of $r^2_{Y(1.2)} = (.5451)^2 = .2971$, we have

$$F = \frac{.2971}{(1 - .6571)/17} = 14.73$$

Note that the F test of $r^2_{Y(1.2)}$ is equivalent to the F test of the regression sum of squares for X_1 when it is entered last (after X_2) in the regression equation. The F test for $r^2_{Y(1.2)}$ is also equal to the square of the t test for b_1.

Computer output most often simply reports the t (or $F = t^2$) test for regression coefficients rather than the equivalent tests of squared semipartial correlation coefficients or of regression sums of squares for variables when they are entered last in the regression equation. It is important, therefore, to know that these tests are all equivalent. If the t (or F) test of a regression coefficient for a given X variable is significant, this means that when this variable is entered *last* in the regression equation, there is a significant increase in the regression sum of squares.

Assume, for example, that we have $k = 5$ X variables and that the test of b_3 is significant. Then

$$r^2_{Y(3.1245)} = R^2_{Y.12345} - R^2_{Y.1245}$$

and

$$F = \frac{r^2_{Y(3.1245)}}{(1 - R^2_{Y.12345})/(n - k - 1)} = \frac{R^2_{Y.12345} - R^2_{Y.1245}}{(1 - R^2_{Y1.2345})/(n - k - 1)}$$

will also be significant. Thus we can also say that $R^2_{Y.12345}$ is significantly larger than $R^2_{Y.1245}$ or that the inclusion of X_3 in the regression equation *after* the other four variables have been entered results in a significant increase in the regression sum of squares. $SS_{tot}R^2_{Y.12345}$ will be the regression sum of squares when all five X variables are in the regression equation. $SS_{tot}R^2_{Y.1245}$ will be the regression sum of squares when only X_1, X_2, X_4, and X_5 are in the regression equation. The difference between these two sums of squares will be equal to the regression sum of squares for X_3 when it is entered last in the regression equation, that is,

$$SS_{tot}r^2_{Y(3.1245)} = SS_{tot}R^2_{Y.12345} - SS_{tot}R^2_{Y.1245}$$

When the X variables are correlated, as X_1 and X_2 are in our example, there is simply no satisfactory method of determining the relative contributions of the X variables to the regression sum of squares or the proportion of $\Sigma(Y - \overline{Y})^2$ accounted for by each of the X variables. This will depend, as we have just seen, on the order in which the X variables are entered in the regression equation. When X_1 is the first variable entered, it accounts for $(.70)^2$ or 49 percent of $\Sigma(Y - \overline{Y})^2$. But if X_1 is entered after X_2, then X_1 accounts for only $(.5451)^2$ or 29.71 percent of $\Sigma(Y - \overline{Y})^2$. Similarly, if X_2 is the first variable entered, it accounts for $(.60)^2$ or 36 percent of $\Sigma(Y - \overline{Y})^2$. But if X_2 is entered after X_1, then X_2 will account for only $(.4088)^2$ or 16.71 percent of $\Sigma(Y - \overline{Y})^2$.

When the X variables are correlated, each variable usually accounts for a larger proportion of $\Sigma(Y - \overline{Y})^2$ when it is entered first in the regression equation than when it follows some other variable or variables. It might seem that if we obtain the regression sum of squares for each X variable when it is entered last in the regression equation, that the sum of these sums of squares would add up to the total regression sum of squares. But that is not true when the X variables are correlated. In our example, the regression sum of squares for X_1 when entered last is 361.31 and the regression sum of squares for X_2 when entered last is 203.23. The sum of these two regression sums of squares is $361.31 + 203.23 = 564.54$, whereas the total regression sum of squares when both X_1 and X_2 are in the regression equation is 799.07.

  When X variables are mutually orthogonal or uncorrelated, the situation is much different, as we shall see in the next section.

9.9 Multiple Correlation When the X Variables Are Mutually Orthogonal

It is important to note, from (9.18) and (9.20), that if the two X variables are uncorrelated, that is, if $r_{12} = 0$, then

$$r_{Y(2.1)} = r_{Y2}$$

and

$$r_{Y(1.2)} = r_{Y1}$$

Thus, if r_{12} is equal to zero, then

$$R^2_{Y.12} = r^2_{Y1} + r^2_{Y2}$$

or simply the sum of the squared correlations of X_1 and X_2 with Y. Similar considerations apply to any number of mutually orthogonal or uncorrelated X variables. If k is the number of X variables and if they are mutually orthogonal or uncorrelated, then

$$R^2_{YY'} = r^2_{Y1} + r^2_{Y2} + \cdots + r^2_{Yk}$$

In this case the contribution of each X variable to the value of $R^2_{YY'}$ is unique and independent of the contribution of every other variable. No matter in which order or in what combination the X variables are entered into the regression equation, the values of r^2_{Yi} will give the proportion of the total sum of squares, $\Sigma(Y - \overline{Y})^2$, that can be accounted for by X_i.[6] Any variable that accounts for a given proportion of the total sum of squares when entered first in the regression equation will also account for the same proportion of the total sum of squares when it is entered last in the regression equation.

---

[6]It is also true, with uncorrelated X variables, that the regression coefficient $b_i = \Sigma x_i y / \Sigma x_i^2$ and the regression sum of squares $SS_{reg} = (\Sigma x_i y)^2 / \Sigma x_i^2$ for any X variable will remain exactly the same no matter in which order or in what combination the X variables are entered into the regression equation.

Exercises

9.1 Assume that we have $n = 20$ observations with

$$\overline{X}_1 = 4.0 \qquad \Sigma x_1^2 = 76.0 \qquad \Sigma x_1 y = 212.8$$
$$\overline{X}_2 = 8.0 \qquad \Sigma x_2^2 = 304.0 \qquad \Sigma x_2 y = 364.8$$
$$\overline{Y} = 20.0 \qquad \Sigma y^2 = 1216.0 \qquad \Sigma x_1 x_2 = 0$$

(a) Find the values of b_1 and b_2 in the multiple regression equation

$$Y' = a + b_1 X_1 + b_2 X_2$$

(b) Find the value of b_1 in the regression equation $Y' = a + b_1 X_1$. Is the value of b_1 equal to the value of b_1 obtained in the multiple regression equation? (c) Find the value of b_2 in the regression equation $Y' = a + b_2 X_2$. Is this value of b_2 equal to the value of b_2 obtained in the multiple regression equation? (d) Find the value of $R_{YY'}^2$. (e) Is $R_{YY'}^2$ significant with $\alpha = .05$? (f) What proportion of the total sum of squares is accounted for by X_1? (g) What proportion of the total sum of squares is accounted for by X_2? (h) What proportion of the total sum of squares is accounted for by X_1 and X_2?

9.2 Assume that we have $n = 20$ observations with

$$\overline{X}_1 = 4.0 \qquad \Sigma x_1^2 = 76.0 \qquad \Sigma x_1 y = 152.0$$
$$\overline{X}_2 = 8.0 \qquad \Sigma x_2^2 = 304.0 \qquad \Sigma x_2 y = 304.0$$
$$\overline{Y} = 20.0 \qquad \Sigma y^2 = 1216.0 \qquad \Sigma x_1 x_2 = 76.0$$

(a) Find the values of r_{Y1} and r_{Y2}. (b) Are these values of r significant with $\alpha = .05$? (c) Can we conclude that b_1 in the regression equation $Y' = a + b_1 X_1$ is significantly different from zero? Explain why or why not. (d) Can we conclude that b_2 in the regression equation $Y' = a + b_2 X_2$ is significantly different from zero? Explain why or why not. (e) Find the value of $R_{YY'}^2$ in the multiple regression equation $Y' = a + b_1 X_1 + b_2 X_2$. (f) Is $R_{YY'}^2$ significant with $\alpha = .05$? (g) Given that X_1 has been included in the regression equation, does X_2 contribute significantly to the reduction in the residual sum of squares? (h) Given that X_2 has been included in the regression equation, does X_1 contribute significantly to the reduction in the residual sum of squares?

9.3 If the largest absolute correlation between Y and any one of k variables is .50, can $R_{YY'}^2$ be smaller than .50? Explain why or why not.

9.4 If $r_{Y1} = .90$ and $r_{Y2} = .60$, is it possible for r_{12} to be equal to zero? Explain why or why not.

Ten

A Little Bit about Matrices and Matrix Algebra

10.1 Introduction

In the preceding chapter we described a multiple regression problem in which we had only two X variables. You need to know, however, about the general case in which we may have any number of X variables. The simplest approach to the general case is in terms of matrices and matrix algebra. This chapter is intended to provide you with the little bit you need to know about matrices and matrix algebra in order to understand the general case of multiple regression involving any number of X variables.

A *matrix,* usually indicated by a boldface letter or symbol, is simply a rectangular array of letters or symbols. If **X** is an $n \times k$ matrix, this means that **X** has n rows and k columns. The number of rows and columns in a matrix is the order of the matrix. An $n \times 1$ matrix is one which has n rows and 1 column and is commonly referred to as a *column vector.* A $1 \times n$ matrix is one which has 1 row and n columns and is commonly referred to as a *row vector.* Vectors are often, but not necessarily, indicated by a lowercase boldface letter or symbol.

Matrix algebra is simply an algebra that is applicable to matrices. You should have little difficulty in understanding the rules regarding addition, subtraction, and multiplication of matrices. Matrix inversion is more complicated than division in ordinary algebra, but it is not necessary that you be able to calculate the inverse of a matrix in order to understand this chapter.

10.2 Addition and Subtraction of Matrices

Two matrices can be added or subtracted provided they both have the same order. For example, if **A** and **B** are both 2×3 matrices with the elements shown below:

$$A = \begin{bmatrix} 1 & 3 & 2 \\ 3 & 2 & 1 \end{bmatrix} \quad B = \begin{bmatrix} 2 & 1 & 4 \\ 3 & 1 & 2 \end{bmatrix}$$

then

$$A + B = \begin{bmatrix} 1+2 & 3+1 & 2+4 \\ 3+3 & 2+1 & 1+2 \end{bmatrix} = \begin{bmatrix} 3 & 4 & 6 \\ 6 & 3 & 3 \end{bmatrix}$$

and

$$A - B = \begin{bmatrix} 1-2 & 3-1 & 2-4 \\ 3-3 & 2-1 & 1-2 \end{bmatrix} = \begin{bmatrix} -1 & 2 & -2 \\ 0 & 1 & -1 \end{bmatrix}$$

10.3 Matrix Multiplication

Suppose that **A** is a 3×2 matrix and that **B** is a 2×2 matrix with the elements shown below:

$$\mathbf{A} = \begin{bmatrix} 3 & 1 \\ 2 & 2 \\ 1 & 4 \end{bmatrix} \qquad \mathbf{B} = \begin{bmatrix} 4 & 3 \\ 2 & 1 \end{bmatrix}$$

If the matrix product $\mathbf{C} = \mathbf{AB}$ is to exist, then the number of columns in **A** must be equal to the number of rows in **B**. For example, if **A** is a $p \times q$ matrix, then **B** must be a $q \times r$ matrix. The matrix $\mathbf{C} = \mathbf{AB}$, for **A** and **B** shown above, exists because **A** has two columns and **B** has two rows. The matrix product of a $p \times q$ matrix and a $q \times r$ matrix will be a $p \times r$ matrix. Thus, the matrix product for the two matrices shown above will be a 3×2 matrix or

$$\mathbf{AB} = \begin{bmatrix} 3 & 1 \\ 2 & 2 \\ 1 & 4 \end{bmatrix} \begin{bmatrix} 4 & 3 \\ 2 & 1 \end{bmatrix} = \begin{bmatrix} 14 & 10 \\ 12 & 8 \\ 12 & 7 \end{bmatrix} = \mathbf{C}$$

In matrix multiplication each row of the matrix on the left multiplies each column of the matrix on the right. The elements in the first row of the product matrix **C** are obtained by multiplying the two columns of **B** by the first row of **A**. Thus

$$(3)(4) + (1)(2) = 14$$

$$(3)(3) + (1)(1) = 10$$

We then multiply the columns of **B** by the second row of **A** to obtain

$$(2)(4) + (2)(2) = 12$$

$$(2)(3) + (2)(1) = 8$$

and these are the two elements of the second row of the product matrix **C**. Finally, we have

$$(1)(4) + (4)(2) = 12$$

$$(1)(3) + (4)(1) = 7$$

and these are the elements in the last row of **C**.

If **A** is a matrix and c is a scalar (a single real number) then

$$cA \qquad\qquad (10.1)$$

means that every element of **A** is to be multiplied by c. For example, if $c = 1/2$ and **A** is the 2×2 matrix shown below, then

$$c\mathbf{A} = .5 \begin{bmatrix} 1 & 2 \\ 3 & 4 \end{bmatrix} = \begin{bmatrix} .5 & 1 \\ 1.5 & 2 \end{bmatrix}$$

10.4　The Matrix Product X'X

For simplicity assume that we have $n = 3$ subjects and one X variable. We let X_0 be an $n \times 1$ *column unit vector* in which all the elements are equal to 1 and X_1 be an $n \times 1$ column vector in which the elements are the values of X associated with each of the $n = 3$ subjects. Then, consider the matrix consisting of the two X vectors:

$$X = \begin{bmatrix} 1 & 3 \\ 1 & 2 \\ 1 & 4 \end{bmatrix}$$

The *transpose* of X, usually indicated by X', will be a matrix in which the rows of X are written as columns in X'. Thus, we have

$$X' = \begin{bmatrix} 1 & 1 & 1 \\ 3 & 2 & 4 \end{bmatrix}$$

The matrix product $X'X$ exists because the number of columns in X' will always be equal to the number of rows in X. For the matrix product in our example we have

$$X'X = \begin{bmatrix} 1 & 1 & 1 \\ 3 & 2 & 4 \end{bmatrix} \begin{bmatrix} 1 & 3 \\ 1 & 2 \\ 1 & 4 \end{bmatrix} = \begin{bmatrix} 3 & 9 \\ 9 & 29 \end{bmatrix}$$

Note that

$$X'X = \begin{bmatrix} 3 & 9 \\ 9 & 29 \end{bmatrix} = \begin{bmatrix} n & \Sigma X_1 \\ \Sigma X_1 & \Sigma X_1^2 \end{bmatrix} \tag{10.2}$$

Assume now that we have n subjects and $k = 3$ X variables. Then the matrix X will consist of the $n \times 1$ unit column vector, X_0, and the $k = 3$ column vectors, X_1, X_2, and X_3, corresponding to the values of the three X variables. Then we have

$$X'X = \begin{bmatrix} n & \Sigma X_1 & \Sigma X_2 & \Sigma X_3 \\ \Sigma X_1 & \Sigma X_1^2 & \Sigma X_1 X_2 & \Sigma X_1 X_3 \\ \Sigma X_2 & \Sigma X_2 X_1 & \Sigma X_2^2 & \Sigma X_2 X_3 \\ \Sigma X_3 & \Sigma X_3 X_1 & \Sigma X_3 X_2 & \Sigma X_3^2 \end{bmatrix} \tag{10.3}$$

We know that $\Sigma X_i X_j = \Sigma X_j X_i$ and $X'X$ is described as a symmetric matrix. A *symmetric* matrix is always a square matrix which has the same elements above the principal diagonal as below the principal diagonal, except that the subscripts are reversed. The *principal diagonal* of a square matrix consists of the elements in the upper left to lower right diagonal.

0.5 The Inverse of a Matrix

The inverse of a square matrix, usually indicated by A^{-1}, if it exists,[1] is a matrix such that the matrix product $A^{-1}A = AA^{-1} = I$. I is called the *identity* matrix and is always a square matrix in which the elements in the principal diagonal are all 1s and all other elements are 0s.[2] An example of a 3×3 identity matrix is shown below:

$$I = \begin{bmatrix} 1 & 0 & 0 \\ 0 & 1 & 0 \\ 0 & 0 & 1 \end{bmatrix}$$

It is obvious that the inverse of an identity matrix is the identity matrix itself. For example,

$$I^{-1}I = \begin{bmatrix} 1 & 0 & 0 \\ 0 & 1 & 0 \\ 0 & 0 & 1 \end{bmatrix}\begin{bmatrix} 1 & 0 & 0 \\ 0 & 1 & 0 \\ 0 & 0 & 1 \end{bmatrix} = \begin{bmatrix} 1 & 0 & 0 \\ 0 & 1 & 0 \\ 0 & 0 & 1 \end{bmatrix} = I$$

Note also that if I is an $n \times n$ matrix and if b is an $n \times 1$ column vector, then

$$Ib = b \tag{10.4}$$

For example,

$$Ib = \begin{bmatrix} 1 & 0 & 0 \\ 0 & 1 & 0 \\ 0 & 0 & 1 \end{bmatrix}\begin{bmatrix} a \\ b_1 \\ b_2 \end{bmatrix} = \begin{bmatrix} a \\ b_1 \\ b_2 \end{bmatrix} = b$$

A *diagonal* matrix is a square matrix in which all of the elements are equal to zero except for those on the principal diagonal. The inverse of a diagonal matrix is easily obtained. For example, if

$$D = \begin{bmatrix} 2 & 0 & 0 \\ 0 & 4 & 0 \\ 0 & 0 & 3 \end{bmatrix} \quad \text{then} \quad D^{-1} = \begin{bmatrix} 1/2 & 0 & 0 \\ 0 & 1/4 & 0 \\ 0 & 0 & 1/3 \end{bmatrix}$$

and $D^{-1}D = DD^{-1} = I$.

The inverse of a 2×2 matrix can also be easily calculated. Assume, for example, that A is a 2×2 matrix with the elements shown below:

$$A = \begin{bmatrix} a & b \\ c & d \end{bmatrix} = \begin{bmatrix} 1 & 2 \\ 3 & 2 \end{bmatrix}$$

[1]The inverse of a square matrix will not exist if the determinant of the matrix is equal to zero.
[2]In algebra for any real number a we have $a^{-1} = 1/a$ and a^{-1} is simply that number for which the product $a^{-1}a = aa^{-1} = 1$. Similarly, in matrix algebra for any square matrix A, A^{-1} is simply that matrix for which the matrix product $A^{-1}A = AA^{-1} = I$.

We first calculate

$$k = \frac{1}{ad - bc} = \frac{1}{(1)(2) - (2)(3)} = -.25$$

Then

$$\mathbf{A}^{-1} = k \begin{bmatrix} d & -b \\ -c & a \end{bmatrix} \tag{10.5}$$

or, in our example,

$$\mathbf{A}^{-1} = -.25 \begin{bmatrix} 2 & -2 \\ -3 & 1 \end{bmatrix} = \begin{bmatrix} -.50 & .50 \\ .75 & -.25 \end{bmatrix}$$

Note now that

$$\mathbf{A}^{-1}\mathbf{A} = \begin{bmatrix} -.50 & .50 \\ .75 & -.25 \end{bmatrix} \begin{bmatrix} 1 & 2 \\ 3 & 2 \end{bmatrix} = \begin{bmatrix} 1 & 0 \\ 0 & 1 \end{bmatrix} = \mathbf{I}$$

10.6 Least Squares Solution in Multiple Regression

For simplicity assume that we have values of a dependent variable Y for each of n subjects and $k = 3$ X variables. Then

$$Y_1 = a + b_1 X_{11} + b_2 X_{12} + b_3 X_{13} + e_1$$
$$Y_2 = a + b_1 X_{21} + b_2 X_{22} + b_3 X_{23} + e_2$$
$$Y_3 = a + b_1 X_{31} + b_2 X_{32} + b_3 X_{33} + e_3$$
$$\cdots\cdots\cdots\cdots\cdots\cdots\cdots\cdots\cdots\cdots\cdots\cdots\cdots\cdots\cdots\cdots$$
$$Y_n = a + b_1 X_{n1} + b_2 X_{n2} + b_3 X_{n3} + e_n$$

We can represent all of these equations compactly in matrix form. For example,

$$\mathbf{Y} = \mathbf{Xb} + \mathbf{e} \tag{10.6}$$

where $\mathbf{Y}$ is an $n \times 1$ column vector of the Y values and $\mathbf{X}$ is an $n \times 4$ matrix or

$$\mathbf{X} = \begin{bmatrix} 1 & X_{11} & X_{12} & X_{13} \\ 1 & X_{21} & X_{22} & X_{23} \\ 1 & X_{31} & X_{32} & X_{33} \\ \cdots & \cdots & \cdots & \cdots \\ 1_n & X_{n1} & X_{n2} & X_{n3} \end{bmatrix}$$

and **b** is a 4×1 column vector of the unknown values of a, b_1, b_2, and b_3 or

$$\mathbf{b} = \begin{bmatrix} a \\ b_1 \\ b_2 \\ b_3 \end{bmatrix}$$

and **e** is an $n \times 1$ column vector of random errors. Then, if we let $\hat{\mathbf{Y}}$ be a column vector of predicted values of Y, as obtained from the regression equation, we have $\hat{\mathbf{Y}} = \mathbf{Xb}$ and

$$\mathbf{Y} - \hat{\mathbf{Y}} = \mathbf{Y} - \mathbf{Xb} = \mathbf{e} \tag{10.7}$$

If the values of a, b_1, b_2, and b_3 are to be those which will minimize $\Sigma(Y - \hat{Y})^2 = \mathbf{e'e} = (\mathbf{Y} - \mathbf{Xb})'(\mathbf{Y} - \mathbf{Xb})$, then it can be shown that they must satisfy the normal equations:

$$
\begin{aligned}
an + b_1\Sigma X_1 &+ b_2\Sigma X_2 &+ b_3\Sigma X_3 &= \Sigma Y \\
a\Sigma X_1 + b_1\Sigma X_1^2 &+ b_2\Sigma X_1 X_2 + b_3\Sigma X_1 X_3 &&= \Sigma X_1 Y \\
a\Sigma X_2 + b_1\Sigma X_2 X_1 &+ b_2\Sigma X_2^2 &+ b_3\Sigma X_2 X_3 &= \Sigma X_2 Y \\
a\Sigma X_3 + b_1\Sigma X_3 X_1 &+ b_2\Sigma X_3 X_2 + b_3\Sigma X_3 &&= \Sigma X_3 Y
\end{aligned}
$$

Note that $\mathbf{X'X}$ will be equal to (10.3) and that for $\mathbf{X'Y}$ we have

$$\mathbf{X'Y} = \begin{bmatrix} \Sigma Y \\ \Sigma X_1 Y \\ \Sigma X_2 Y \\ \Sigma X_3 Y \end{bmatrix} \tag{10.8}$$

Consequently, the normal equations can be expressed in matrix notation as

$$\mathbf{X'Xb} = \mathbf{X'Y} \tag{10.9}$$

To solve for the unknown values of **b**, we first find $(\mathbf{X'X})^{-1}$. Then multiplying both sides of (10.9) by $(\mathbf{X'X})^{-1}$ we have

$$(\mathbf{X'X})^{-1}(\mathbf{X'X})\mathbf{b} = (\mathbf{X'X})^{-1}\mathbf{X'Y} \tag{10.10}$$

but

$$(\mathbf{X'X})^{-1}(\mathbf{X'X}) = \mathbf{I}$$

and, as (10.4) shows,

$$\mathbf{Ib} = \mathbf{b}$$

Consequently, the unknown values of **b** will be given by

$$\mathbf{b} = (\mathbf{X'X})^{-1}\mathbf{X'Y} \tag{10.12}$$

The solution shown is perfectly general and applies no matter how many different X variables might be involved. In order to obtain a solution for the unknown values of **b**, it is only necessary that the inverse of **X'X** exists.[3] There are routine arithmetic methods by which $(X'X)^{-1}$ can be obtained, assuming that it exists, but the methods are extremely tedious and time consuming and particularly so if the number of X variables is large. The solution for $(X'X)^{-1}$, on the other hand, is quickly and easily accomplished by high-speed electronic computers.

In all the examples discussed in the following chapters and in the exercises for these chapters the inverse, $(X'X)^{-1}$, will be given. The only exception will be for the inverse of a 2×2 matrix or for a diagonal matrix. You can easily calculate the inverse for these matrices.

Exercises

10.1   For which of the pairs of matrices listed below does the product **C** = **AB** exist? If **C** = **AB** exists, what is the order of **C**?
(a) **A**: 3×4 **B**: 4×3
(b) **A**: 2×2 **B**: 2×2
(c) **A**: 1×4 **B**: 4×1
(d) **A**: 5×5 **B**: 5×5
(e) **A**: 6×2 **B**: 6×2
(f) **A**: 4×1 **B**: 1×4
(g) **A**: 4×3 **B**: 3×4

10.2   For which of the matrices in Exercise 10.1 does the product **C** = **BA** exist? If the product **C** = **BA** exists, what is the order of **C**?
10.3   For which of the matrices in Exercise 10.1 can we add the two matrices to obtain **C** = **A** + **B**?
10.4   For the matrix **X** shown below write the transpose of **X**.

$$X = \begin{bmatrix} 1 & 2 & 3 \\ 2 & 1 & 2 \\ 3 & 0 & 0 \\ 0 & 1 & 1 \end{bmatrix}$$

10.5   For the **X** matrix in Exercise 10.4 write the product **X'X**. Is **X'X** symmetric? If so, explain why.
10.6   For the matrix **X** in Exercise 10.4, does the product **XX'** exist? If so, explain why. Is **XX'** equal to **X'X**?

---

[3]The inverse of **X'X** will not exist if $r_{ij}^2 = 1$ for any two of the X variables or if $R_{i.}^2$, the squared multiple correlation of X_i with the other X variables, is equal to one. In either of these cases, the determinant of **X'X** will be equal to zero.

10.7 (a) Is the **X** matrix shown below symmetric? If so, explain why.

$$\mathbf{X} = \begin{bmatrix} 1 & 1 & 2 \\ 1 & 2 & 0 \\ 2 & 0 & 3 \end{bmatrix}$$

(b) Write the transpose of **X**. Is **X'** = **X**?

10.8 (a) What is the inverse of the matrix **X** shown below?

$$\mathbf{X} = \begin{bmatrix} 4 & 0 & 0 \\ 0 & 2 & 0 \\ 0 & 0 & 1 \end{bmatrix}$$

(b) Show that $\mathbf{X}^{-1}\mathbf{X} = \mathbf{X}\mathbf{X}^{-1} = \mathbf{I}$.

10.9 Suppose we have $n = 5$ values of Y and one X variable with the values shown below:

Y	X_0	X_1
1	1	2
2	1	6
3	1	4
4	1	10
5	1	8

(a) Calculate the values of $a = \overline{Y} - b\overline{X}$ and $b = \Sigma xy/\Sigma x^2$.

(b) Calculate **X'X** and the inverse of **X'X**. Note that **X'X** will be a 2×2 matrix and the inverse can be found using the method described in the chapter for a 2×2 matrix. Calculate $(\mathbf{X'X})^{-1}\mathbf{X'Y} = \mathbf{b}$. The values of a and b should be equal to those obtained in (a).

Eleven

Multiple Regression: The General Case

11.1 Introduction

Table 11.1 gives the values of three X variables and one Y variable for $n = 10$ cases or subjects. The table also shows the unit column vector $\mathbf{X}_0$ in which all the elements are equal to one. We use the data in Table 11.1 to illustrate some of the matrix calculations involved in a multiple regression analysis. Although we have only three X variables, the calculations apply to the general case where we may have any number of X variables.

11.2 Calculation of b

With the unit vector $\mathbf{X}_0$ and $k = 3$ X variables, we have the 4×4 symmetric matrix $\mathbf{X'X}$ or[1]

$$\mathbf{X'X} = \begin{bmatrix} n & \Sigma X_1 & \Sigma X_2 & \Sigma X_3 \\ \Sigma X_1 & \Sigma X_1^2 & \Sigma X_1 X_2 & \Sigma X_1 X_3 \\ \Sigma X_2 & \Sigma X_2 X_1 & \Sigma X_2^2 & \Sigma X_2 X_3 \\ \Sigma X_3 & \Sigma X_3 X_1 & \Sigma X_3 X_2 & \Sigma X_3^2 \end{bmatrix} = \begin{bmatrix} 10 & 53 & 56 & 53 \\ 53 & 345 & 323 & 299 \\ 56 & 323 & 404 & 289 \\ 53 & 299 & 289 & 351 \end{bmatrix}$$

We also have the 4×1 column vector $\mathbf{X'Y}$ or

$$\mathbf{X'Y} = \begin{bmatrix} \Sigma Y \\ \Sigma X_1 Y \\ \Sigma X_2 Y \\ \Sigma X_3 Y \end{bmatrix} = \begin{bmatrix} 53 \\ 317 \\ 332 \\ 290 \end{bmatrix}$$

TABLE 11.1   Values of Y and of $k = 3$ X Variables for $n = 10$ Subjects

Y	X_0	X_1	X_2	X_3
3	1	4	1	8
2	1	1	2	1
5	1	8	3	9
7	1	7	4	3
4	1	2	5	5
7	1	5	10	3
7	1	8	6	4
4	1	3	7	7
9	1	7	8	9
5	1	8	10	4

[1]Note that $\mathbf{X'X}$ contains all the values needed for calculating the mean and variance of each of the X variables and also the covariance between any two X variables.

If we can find the inverse of the $X'X$ matrix, which we designate by $(X'X)^{-1}$, then it can be shown that

$$(X'X)^{-1}X'Y = b \tag{11.1}$$

where b is a column vector containing the least squares values of a, b_1, b_2, and b_3. The inverse of $X'X$ is defined if $(X'X)^{-1}X'X = I$, where I is an identity matrix with 1s in the upper left to lower right diagonal and 0s elsewhere or, in the case of 4×4 matrix, if

$$(X'X)^{-1}X'X = \begin{bmatrix} 1 & 0 & 0 & 0 \\ 0 & 1 & 0 & 0 \\ 0 & 0 & 1 & 0 \\ 0 & 0 & 0 & 1 \end{bmatrix} = I$$

In the example under discussion the inverse of the $X'X$ matrix can be shown to be

$$(X'X)^{-1} = \begin{bmatrix} 1.006289 & -.039200 & -.056781 & -.071803 \\ -.039200 & .019788 & -.006236 & -.005803 \\ -.056781 & -.006236 & .013134 & .003072 \\ -.071803 & -.005803 & .003072 & .016105 \end{bmatrix}$$

We have

$$X'Y - \begin{bmatrix} 53 \\ 317 \\ 332 \\ 290 \end{bmatrix}$$

Then

$$(X'X)^{-1}X'Y = b = \begin{bmatrix} a \\ b_1 \\ b_2 \\ b_3 \end{bmatrix}$$

where

$$a = (1.006289)(53) + (-.039200)(317)$$
$$\qquad + (-.056781)(332) + (-.071803)(290)$$
$$= 1.2318$$
$$b_1 = (-.039200)(53) + (.019788)(317)$$
$$\qquad + (-.006236)(332) + (-.005803)(290)$$
$$= .4420$$

$$b_2 = (-.056781)(53) + (-.006236)(317)$$
$$+ (.013134)(332) + (.003072)(290)$$
$$= .2652$$
$$b_3 = (-.071803)(53) + (-.005803)(317)$$
$$+ (.003072)(332) + (.016105)(290)$$
$$= .0452$$

Then we have the regression equation

$$Y' = 1.233 + .442X_1 + .265X_2 + .045X_3$$

11.3 Calculation of SS_{reg} and SS_{res}

To find the regression sum of squares, we first calculate

$$\mathbf{b'X'Y} = (1.2328)(53) + (.4420)(317) + (.2652)(332) + (.0452)(290)$$
$$= 306.6068$$

Then, the regression sum of squares will be given by

$$SS_{reg} = \mathbf{b'X'Y} - \frac{(\Sigma Y)^2}{n} \tag{11.2}$$

For the Y values in Table 11.1 we have $\Sigma Y = 53$. Then, using (11.2), we obtain

$$SS_{reg} = 306.6068 - \frac{(53)^2}{10} = 25.7068$$

For the Y values in Table 11.1 we also have $SS_{tot} = \Sigma(Y - \overline{Y})^2 = 42.1000$. Then the residual sum of squares will be equal to

$$SS_{res} = SS_{tot} - SS_{reg} = 42.1000 - 25.7068 = 16.3932$$

11.4 Calculating $R_{YY'}^2$ and a Test of Significance of $R_{YY'}^2$

We know that

$$R_{YY'}^2 = \frac{SS_{reg}}{SS_{tot}}$$

and, in our example, we have

$$R_{Y.123}^2 = \frac{25.7068}{42.1000} = .6106$$

For a test of significance of $R^2_{Y.123}$, we have

$$F = \frac{.6106/3}{(1 - .6106)/6} = 3.14$$

which is not a significant value of F with $\alpha = .05$ and with 3 and 6 d.f. Equivalently, we have

$$F = \frac{MS_{reg}}{MS_{res}} = \frac{25.7068/3}{16.3932/6} = 3.14$$

11.5 Standard Errors of the Regression Coefficients

The standard error of a regression coefficient will be given by

$$s_{b_i} = \sqrt{MS_{res} c_{ii}} \tag{11.3}$$

where MS_{res} is the residual mean square and c_{ii} is the diagonal element of $(X'X)^{-1}$ corresponding to b_i.

In our example, the three standard errors will be equal to[2]:

$$s_{b_1} = \sqrt{(2.7322)(.019788)} = .2325$$
$$s_{b_2} = \sqrt{(2.7322)(.013134)} = .1894$$
$$s_{b_3} = \sqrt{(2.7322)(.016105)} = .2098$$

The standard errors of the regression coefficients are shown in Table 11.2 along with the regression coefficients and the t tests of significance of each of the regression coefficients. None of the regression coefficients is significant with $\alpha = .05$.

TABLE 11.2   Values of b_i, s_{b_i}, and the t Test for b_i

Variable	b_i	s_{b_i}	t
X_1	.4420	.2325	1.9011
X_2	.2652	.1894	1.4002
X_3	.0452	.2098	.2154

[2]Observe that a very large value of c_{ii} will also result in a very large value for s_{b_i}. As shown in the next section, a large value of c_{ii} is an indication that X_i has a large multiple correlation with the other X variables.

11.6 Correlations among the X Variables

In multiple correlation and regression problems a large correlation coefficient, r_{ij}, between any two X variables or a large multiple correlation of any one of the X variables with the remaining X variables can cause problems. We let $R_{i.}^2$ be the squared multiple correlation of X_i with the remaining X variables. Then if $R_{i.}^2 = 1$, the inverse of $\mathbf{X'X}$ does not exist. If $R_{i.}^2$ is large, then the standard error of the corresponding regression coefficient b_i for that variable will also be large. The standard error of b_i in a multiple regression problem will also be given by

$$s_{b_i} = \sqrt{\frac{MS_{res}}{\Sigma x_i^2(1 - R_{i.}^2)}} \tag{11.4}$$

and, obviously, as $R_{i.}^2$ becomes large the denominator of (11.4) becomes small and s_{b_i} becomes large.

If the inverse $(\mathbf{X'X})^{-1}$ is available, then it is relatively easy to obtain $R_{i.}^2$ for any of the X variables. For example,

$$R_{i.}^2 = 1 - \frac{1}{c_{ii}(n-1)s_i^2} \tag{11.5}$$

where c_{ii} is the diagonal element of $(\mathbf{X'X})^{-1}$ corresponding to X_i and s_i^2 is the variance of X_i. The variance s_i^2, for any X_i, can easily be calculated from the values in $\mathbf{X'X}$. If we want the squared multiple correlation of X_1 with X_2 and X_3, in our example, then we have $c_{11} = .019788$, $n = 10$, and $s_1 = 2.67$. Then substituting in (11.5), we have

$$R_{1.}^2 = 1 - \frac{1}{(.019788)(10-1)(2.67)^2} = .2124 = R_{1.23}^2$$

If $(\mathbf{X'X})^{-1}$ is not available, we can obtain $R_{i.}^2$ if we know MS_{res}, s_{b_i}, the standard error of b_i, and the standard deviation s_i of X_i. In this instance,

$$R_{i.}^2 = 1 - \frac{MS_{res}}{s_{b_i}^2(n-1)s_i^2} \tag{11.6}$$

For X_1 we have $s_1 = 2.67$, and $s_{b_1} = .2325$, and $MS_{res} = 2.7322$. Then, substituting in (11.6), we have

$$R_{1.}^2 = 1 - \frac{2.7322}{(.2325)^2(10-1)(2.67)^2} = .2122$$

which is equal within rounding errors to the value we obtained before.

11.7 Adjusted $R_{YY'}^2$

It is always possible to predict perfectly n values of a dependent variable Y from a set of $k = n - 1$ X variables. That is one of the reasons why, in general,

n, the number of observations in a multiple regression problem, should be considerably larger than k, the number of X variables.

Because $R^2_{YY'}$ is dependent on both n and k, it is often useful to calculate the value of $R^2_{YY'}$ adjusted for degrees of freedom. A commonly used formula for the adjusted value is

$$\text{Adjusted } R^2_{YY'} = 1 - (1 - R^2_{YY'})\frac{n-1}{n-k-1} \tag{11.7}$$

In the example we have been discussing, we have $R^2_{YY'} = .6106$ with $n = 10$ and $k = 3$. Then, for the adjusted value of $R^2_{YY'}$, we have

$$\text{Adjusted } R^2_{YY'} = 1 - (1 - .6106)\frac{10-1}{10-3-1} = .4159$$

a value which is considerably smaller than the obtained value of $R^2_{YY'} = .6106$.

11.8 Package Programs and Computer Output

A multiple regression analysis involving three or more X variables and a relatively large number of observations *can* be accomplished with a hand calculator but the calculations are extremely tedious, whereas the necessary calculations are quickly and easily accomplished by modern computers. Almost all college and university computer centers have one or more package programs which will do a multiple regression analysis. In Figure 11.1 we show part of the computer output obtained with one package program, *MINITAB*, for the example discussed in the chapter involving three X variables and $n = 10$ observations.

The first line in Figure 11.1 gives the regression equation. The least squares values of a, b_1, b_2, and b_3 are then shown along with their standard errors and the corresponding t tests.[3] Note that *MINITAB* has rounded the t values to two decimal places, whereas we reported the t values to four decimal places in Table 11.2. The standard deviation of Y about the regression line, $s_{Y.X} = 1.653$, is simply the square root of $MS_{res} = 2.733$. The output then shows the obtained value of $R^2_{YY'} = .611$ and the value of $R^2_{YY'}$ adjusted for degrees of freedom. The adjusted value of $R^2_{YY'} = .416$ is based on equation (11.7) given earlier in the chapter.

The analysis of variance provides the regression sum of squares and the residual sum of squares and the corresponding mean squares. The output also gives the regression sum of squares for the X variables when they are entered in the sequence X_1, X_2, and X_3.

---

[3]The standard error of the Y-intercept will be given by $\sqrt{MS_{res}c_{00}}$, where c_{00} is the first diagonal element in $(\mathbf{X'X})^{-1}$.

```
THE REGRESSION EQUATION IS

Y = 1.23 + .442X1 + .265X2 + .0452X3

                                   ST. DEV.      T-RATIO =
       COLUMN     COEFFICIENT      OF COEF.      COEF/S.D.

         —          1.233          1.658          0.74
         X1         0.4420         0.2325         1.90
         X2         0.2652         0.1895         1.40
         X3         0.0452         0.2098         0.22

THE ST. DEV. OF Y ABOUT REGRESSION LINE IS

S = 1.653

WITH (10 — 4) = 6 DEGREES OF FREEDOM

R-SQUARED = 61.1 PERCENT
R-SQUARED = 41. PERCENT, ADJUSTED FOR D.F.

ANALYSIS OF VARIANCE

       DUE TO        DF           SS          MS = SS/DF
     REGRESSION       3         25.703          8.568
     RESIDUAL         6         16.397          2.733
     TOTAL            9         42.100

FURTHER ANALYSIS OF VARIANCE
SS EXPLAINED BY EACH VARIABLE WHEN ENTERED IN THE
ORDER GIVEN

       DUE TO        DF           SS

     REGRESSION       3         25.703
     X1               1         20.331
     X2               1          5.245
     X3               1          0.127
```

FIGURE 11.1   Partial computer output obtained with *MINITAB*.

Although the *MINITAB* output also provides the inverse $(X'X)^{-1}$, we do not show this in Figure 11.1 because the inverse has already been reported earlier in the chapter.

11.9 Semipartial Correlations

From Figure 11.1 we see that the regression sum of squares for X_3 when it is entered *last* in the regression equation is .127. If we divide this sum of squares by $SS_{tot} = 42.100$, we will obtain $r^2_{Y(3.12)} = .127/42.100 = .0030$. The squared semipartial correlation coefficient $r^2_{Y(3.12)}$ is simply the proportion of the total sum of squares accounted for by X_3 when it is entered last in the regression equation.

The F test of significance for $r^2_{Y(3.12)}$ can be shown to be equal to the square of the t test for b_3. For example, we have

$$F = \frac{.0030}{(1 - .611)/6} = \frac{.0030}{.0648} = .0463$$

and $t = \sqrt{.0463} = .22$ and this is the rounded value of the t test for the significance of b_3.

Then we also have

$$r^2_{Y(3.12)} = (.22)^2 \frac{1 - .611}{6} = .0031$$

Although the values of $r^2_{Y(1.23)}$ and $r^2_{Y(2.13)}$ are not part of the computer output, they can obviously be obtained from the t tests of b_1 and b_2, respectively, in the same manner in which we obtained $r^2_{Y(3.12)}$ from the t test of b_3. Thus, we have

$$r^2_{Y(1.23)} = (1.90)^2 \frac{1 - .611}{6} = .2339$$

and

$$r^2_{Y(2.13)} = (1.40)^2 \frac{1 - .611}{6} = .1270$$

We observe from Figure 11.1 that when X_1 is entered *first* in the regression equation it accounts for $r^2_{Y1} = 20.331/42.100 = .4829$ of the total sum of squares, whereas when X_1 is entered *last* in the regression equation it accounts for only .2339 of the total sum of squares. As we have emphasized previously, when the X variables are correlated, the order of their entry into the regression equation makes a considerable difference in the proportion of the total sum of squares accounted for by each variable. It is only when the X variables are uncorrelated with each other that each will have the same regression sum of squares no matter in which order they are entered into the regression equation.

If we divide the regression sums of squares for X_1, X_2, and X_3, as given in Figure 11.1, by $SS_{tot} = 42.100$, we obtain

$$r^2_{Y1} = .4829 \qquad r^2_{Y(2.1)} = .1246 \qquad r^2_{Y(3.12)} = .0030$$

Then, we also have

$$R^2_{Y.123} = r^2_{Y1} + r^2_{Y(2.1)} + r^2_{Y(3.12)} \tag{11.8}$$

Equation (11.8) is perfectly general. For example, with $k = 4$ X variables we would have

$$R^2_{Y.1234} = r^2_{Y1} + r^2_{Y(2.1)} + r^2_{Y(3.12)} + r^2_{Y(4.123)}$$

We also have

$$R^2_{Y.12} = r^2_{Y1} + r^2_{Y(2.1)}$$

so that

$$R^2_{Y.123} - R^2_{Y.12} = r^2_{Y(3.12)} \qquad (11.9)$$

Equation (11.9) is also perfectly general. For example, with $k = 4$ X variables we would have

$$R^2_{Y.1234} - R^2_{Y.123} = r^2_{Y(4.123)}$$

and

$$R^2_{Y.1234} - R^2_{Y.12} = R^2_{Y(34.12)}$$

and $R^2_{Y(34.12)}$ is the proportion of the total sum of squares accounted for X_3 *and* X_4 given that X_1 and X_2 are already in the regression equation.

11.10 Selecting X Variables for Prediction of Y

Suppose that we have $k = 10$ X variables and $n = 200$ observations. It is not unreasonable to believe that the X variables will have varying degrees of correlation with each other and also with the dependent or Y variable. It is also not unreasonable to believe that because the X variables are correlated with each other that a selected set of $k' < k$ of the X variables will result in a value of $R^2_{Yk'}$ that is not much smaller than the value of R^2_{Yk} based on all k variables.

A number of different methods are available in computer package programs for selecting a subset of $k' < k$ variables.[4] One method, called the *forward solution,* is to select first that variable that has the largest absolute correlation with Y. The second variable selected is the one with the highest squared semipartial correlation with Y given that the one X variable already selected is in the regression equation. After two variables have been selected, the next variable selected is the one that has the largest squared semipartial

[4]A more complete discussion of these methods can be found in the manuals for the various package programs. See, for example: T. A. Ryan, B. L. Joiner, and B. F. Ryan, *MINITAB Student Handbook*. North Scituate, Mass.: Duxbury, 1976; N. H. Nie, C. H. Hull, J. G. Jenkins, K. Steinbrenner, and D. H. Bent, *SPSS: Statistical Package for the Social Sciences* (2d ed.). New York: McGraw-Hill, 1975; W. J. Dixon and M. B. Brown, *BMDP-77 Biomedical Computer Programs, P-Series*. Berkeley: University of California Press, 1977.

correlation with Y given that the first two variables selected are already in the regression equation.

Although an F to enter option is available which will terminate the process when a new variable does not exceed the value of F to enter, there is no reason why the process should not be continued until a reasonable number of variables have been selected.[5] It is then possible to determine just how many variables should be used to obtain a given value of $R_{Yk'}^2$. With a large number of observations, even variables with relatively small squared semipartial correlations, that is variables that contribute little to the regression sum of squares, may be selected by the F to enter criterion.

One of the problems with the forward solution is that it does not correct for the fact that a variable that is selected at an early stage may contribute little to the regression sum of squares after a number of additional variables have been selected. It is worthwhile, therefore, to examine the t tests of the regression coefficients after a given set of k' variables have been selected. Recall, as we showed earlier, that the t test of a regression coefficient when k' variables are in the regression equation is equivalent to a test of significance of a given variable when it is entered last in the regression equation. Thus, it is possible that if X_i is selected early in the forward solution, it may not contribute much to the regression sum of squares when entered last in the regression equation.

A *stepwise solution* for selecting k' variables is also available. A stepwise solution is similar to the forward solution in that an F to enter a variable at each step or stage is used. In addition, however, the stepwise solution also uses an F to remove or delete a variable that has been selected at an earlier stage.[6] For example, after the first two variables have been selected, as in the forward solution, an F test is then made to determine the contribution of the variable first selected when it is entered after the second variable selected. Similarly, after three variables have been selected, F tests are made of the first two variables selected to determine their contribution when entered last in the regression equation. The results of these tests are compared with the value of F to remove a variable. Thus, the stepwise solution uses not only a value of F to enter a variable but also another value of F for removing a variable from the regression equation at a later stage.

The *backward solution* for selecting $k' < k$ variables is just the reverse of the forward solution. With the backward solution one starts with the value of R_{Yk}^2, that is, the squared multiple correlation of Y with all k variables in the regression equation. The variable that contributes least to the regression sum

---

[5]The value of F to enter is more or less arbitrary because the tests are not exact tests of significance. Commonly used values of F to enter are 3 and 4.

[6]Again, the value of F to remove is more or less an arbitrary choice. One might choose to remove a variable if the F test for that variable is less than some arbitrary value, say less than 3 or less than 4.

of squares when entered last is then deleted and a new value of the squared multiple correlation with the remaining $k - 1$ variables is calculated. At this stage the variable that contributes least to the regression sum of squares when entered last is deleted and a new value of the squared multiple correlation is calculated using only the remaining $k - 2$ variables. The procedure continues until a set of k' variables have been retained. Again the criterion for deleting variables may be based on a value of F to remove, or one may continue the process until left with a reasonable set of $k' < k$ variables. If the squared multiple correlation at this stage is judged to be too small, then the last variable that was deleted can be retained; if the squared multiple correlation is still judged to be too small, the next to last variable can be included, and so on.

Suppose that we use each of the three methods we have described to select $k' = 5$ variables from a much larger set of k variables, without regard to the value of F to enter or F to remove. In other words, we use each method simply to find the $k' = 5$ variables that would be selected for inclusion in the regression equation. If we do, then, in general, we will find that the set of $k' = 5$ variables selected by each method are not necessarily the same. To find the subset of $k' = 5$ variables that will result in the largest squared multiple correlation with Y, we would need to examine all possible subsets of $k' = 5$ drawn from the larger set of k variables. This means that if $k = 10$, then we would have $10!/5!5! = 252$ subsets of $k' = 5$ variables to examine if we wish to determine which one results in the largest squared multiple correlation with Y.

Exercises

11.1   We have $n = 10$ values of Y and $k = 3$ X variables as shown below.

Y	X_0	X_1	X_2	X_3
4	1	1	5	9
3	1	2	7	4
5	1	2	9	7
8	1	3	11	8
10	1	3	7	7
7	1	4	5	5
12	1	4	9	9
11	1	5	10	12
12	1	6	11	13
14	1	6	14	11

(a) Calculate $\mathbf{X'X}$.

(b) The inverse of $\mathbf{X'X}$ can be shown to be:

$$(\mathbf{X'X})^{-1} = \begin{bmatrix} 1.355996 & .054923 & -.094095 & -.073609 \\ .054923 & .085772 & -.022441 & -.019556 \\ -.094095 & -.022441 & .028869 & -.009314 \\ -.073609 & -.019556 & -.009314 & .026585 \end{bmatrix}$$

Calculate $\mathbf{X'Y}$, and $(\mathbf{X'X})^{-1}\mathbf{X'Y} = \mathbf{b}$.

(c) Calculate SS_{reg} and SS_{res}.

(d) Calculate $R^2_{YY'}$ and test $R^2_{YY'}$ for significance.

(e) Calculate s_{b_i} for each b_i and test each for significance.

(f) Find the squared multiple correlation of X_1 with X_2 and X_3.

Twelve

Multiple Regression with Standardized Variables

12.1 Standardized Variables

Some individuals who work with multiple regression analyses choose to standardize the Y and X variables. A *standardized variable* is one for which the mean is equal to zero and which has variance and standard deviation equal to one. Any variable can be transformed into a standardized variable by subtracting the mean from each value of the variable and dividing the remainder by the standard deviation. We shall refer to standardized variables as z variables.[1] Thus, we have

$$z = \frac{X - \overline{X}}{s} \tag{12.1}$$

and it is easy to see that $\overline{z}$ is equal to zero because

$$\Sigma z = \frac{1}{s} \Sigma (X - \overline{X})$$

and we know that $\Sigma(X - \overline{X}) = 0$.

Because $\overline{z}$ is equal to zero, the variance of z will be

$$s_z^2 = \frac{\Sigma z^2}{n - 1} = \frac{\Sigma (X - \overline{X})^2}{s^2(n - 1)}$$

Because

$$s^2 = \frac{\Sigma (X - \overline{X})^2}{n - 1}$$

we see that the variance of z will be equal to one and consequently the standard deviation of z will also be equal to one.

If all variables are in standardized form, then the Y-intercept or a will be equal to zero because the means of all the variables will be equal to zero. Furthermore, we know that if all the variables are standardized, then the covariance between any two variables,

$$c_{z_i z_j} = \frac{\Sigma z_i z_j}{n - 1} = r_{ij} s_{z_i} s_{z_j} = r_{ij}$$

because $s_{z_i} = s_{z_j} = 1$.

With standardized variables, the regression coefficients will not be equal to those obtained with unstandardized variables. To indicate a standardized regression coefficient, we will use $\tilde{b}_i$. We can obtain the values of the unstandardized regression coefficients from the standardized coefficients

---

[1]The standardized variable z should not be confused with the transformation z, for the correlation coefficient.

because

$$b_i = \frac{s_Y}{s_i} \tilde{b}_i \qquad (12.2)$$

where s_Y is the standard deviation of the Y variable and s_i is the standard deviation of X_i.

12.2 The Regression Equation and $R^2_{YY'}$

With standardized variables, the regression equation becomes

$$y' = \tilde{b}_1 z_1 + \tilde{b}_2 z_2 + \cdots + \tilde{b}_k z_k \qquad (12.3)$$

where $z_1, z_2, \ldots, z_k$ are the standardized X variables. It is easy to see that if we sum (12.3) over all n values, then all terms on the right sum to zero, because the $\tilde{b}_i$'s are constants and $\tilde{b}_i \Sigma z_i = 0$. Consequently, the regression sum of squares

$$SS_{reg} = \Sigma(Y' - \overline{Y})^2$$

with standardized variables will be $\Sigma y'^2$ because $\overline{y}$ is equal to zero. Thus, with standardized variables, we have

$$SS_{reg} = \Sigma y'^2 = \Sigma(\tilde{b}_1 z_1 + \tilde{b}_2 z_2 + \cdots + \tilde{b}_k z_k)^2 \qquad (12.4)$$

Note that if we expand (12.4), sum, and divide by $n - 1$, then all values of $\tilde{b}_i^2 \Sigma z_i^2 / (n - 1)$ will be equal to $\tilde{b}_i^2$ because $\Sigma z_i^2 / (n - 1)$ is equal to one. Similarly, all values of $\tilde{b}_i \tilde{b}_j \Sigma z_i z_j / (n - 1)$ will be equal to $\tilde{b}_i \tilde{b}_j r_{ij}$ because all values of $\Sigma z_i z_j / (n - 1)$ will be equal to r_{ij}. Thus we have

$$\frac{\Sigma y'^2}{n - 1} - \Sigma \tilde{b}_i^2 + 2\Sigma \tilde{b}_i \tilde{b}_j r_{ij} \qquad (12.5)$$

We know that the squared multiple correlation coefficient will be given by

$$R^2_{YY'} = \frac{SS_{reg}}{SS_{tot}}$$

but with standardized variables $SS_{reg} = \Sigma y'^2$ and $SS_{tot} = \Sigma z_y^2 = n - 1$. Consequently, with standardized variables

$$R^2_{YY'} = \frac{\Sigma y'^2}{n - 1} = \Sigma \tilde{b}_i^2 + 2\Sigma \tilde{b}_i \tilde{b}_j r_{ij} \qquad (12.6)$$

or simply the variance of the predicted values.

12.3 Matrix Calculations

With standardized variables, $a = 0$, and we have no need for the unit vector X_0. Then, with $k = 3$ standardized variables, we have a 3×3 symmetric matrix $X'X$ or

$$X'X = \begin{bmatrix} \Sigma z_1^2 & \Sigma z_1 z_2 & \Sigma z_1 z_3 \\ \Sigma z_2 z_1 & \Sigma z_2^2 & \Sigma z_2 z_3 \\ \Sigma z_3 z_1 & \Sigma z_3 z_2 & \Sigma z_3^2 \end{bmatrix}$$

Multiplying $X'X$ by the scalar, $1/(n-1)$, we have the correlation matrix of the X variables, or

$$R = \begin{bmatrix} 1 & r_{12} & r_{13} \\ r_{21} & 1 & r_{23} \\ r_{31} & r_{32} & 1 \end{bmatrix}$$

For the $n = 10$ cases with $k = 3$ X variables in the preceding chapter in Table 11.1, we have[2]

$$R = \begin{bmatrix} 1.00000 & .34418 & .27002 \\ .34418 & 1.00000 & -.09798 \\ .27002 & -.09798 & 1.00000 \end{bmatrix}$$

Similarly, if Y is standardized, then $X'Y[1/(n-1)]$ will be a column vector r_{Yi} which contains the values of r_{Y1}, r_{Y2}, and r_{Y3}. In our example, these can be shown to be equal to .695, .571, and .168, respectively. Then, if we find the inverse of R, the least squares solution for the standardized regression coefficients will be given by

$$R^{-1} r_{Yi} = \tilde{b} \qquad (12.7)$$

where r_{Yi} is the column vector of the correlations of Y with the three X variables and $\tilde{b}$ is a column vector with the values of $\tilde{b}_1$, $\tilde{b}_2$, and $\tilde{b}_3$.

In our example, we have

$$R^{-1} r_{Yi} = \begin{bmatrix} 1.26841 & -.47468 & -.38900 \\ -.47468 & 1.18733 & .24451 \\ -.38900 & .24451 & 1.12900 \end{bmatrix} \begin{bmatrix} .695 \\ .571 \\ .168 \end{bmatrix} = \begin{bmatrix} .54515 \\ .38914 \\ .05893 \end{bmatrix} = \tilde{b}$$

and the standardized regression coefficients are $\tilde{b}_1 = .54515$, $\tilde{b}_2 = .38914$, and $\tilde{b}_3 = .05893$. Then the regression equation will be[3]

$$y' = .54515 z_1 + .38914 z_2 + .05893 z_3$$

---

[2]We use the same example we used in the preceding chapter in order to show the relationships between the two analyses.

[3]The standardized regression coefficient $\tilde{b}_i$ is the amount by which y' will change with a change of one standard deviation in X_i, if all other variables are held constant.

12.4 Calculating $R_{Y.123}^2$

The values of r_{12}, r_{13}, and r_{23}, as shown earlier in **R**, are equal to .34418, .27002, and $-.09798$, respectively. Then, for our example, we have

$$\Sigma \tilde{b}_i^2 = (.54515)^2 + (.38914)^2 + (.05893)^2 = .45209$$

and

$$2\Sigma \tilde{b}_i \tilde{b}_j r_{ij} = 2[(.54515)(.38914)(.34418) + (.54515)(.05893)(.27002)$$
$$+ (.38914)(.05893)(-.09798)]$$
$$= .15889$$

and, substituting in (12.6) we obtain

$$R_{Y.123}^2 = .45209 + .15889 = .611$$

which is equal, within rounding errors, to the value reported in the preceding chapter.

For a test of significance of $R_{Y.123}^2$, we have

$$F = \frac{.611/3}{(1 - .611)/6} = 3.14$$

which is equal, as it should be within rounding errors, to the F test shown in the preceding chapter.

12.5 The Unstandardized Regression Coefficients

To calculate the unstandardized regression coefficients, we use (12.2). Then, we have

$$b_1 = \frac{(2.16)(.54515)}{2.67} = .4410$$

$$b_2 = \frac{(2.16)(.38914)}{3.17} = .2652$$

$$b_3 = \frac{(2.16)(.05893)}{2.79} = .0456$$

which are equal, within rounding errors, to the values we obtained in the preceding chapter.

12.6 Calculating $R_{i.}^2$

The diagonal elements of $\mathbf{R}^{-1}$ can be used to obtain the squared multiple correlation of any one of the X variables with all of the other X variables. For example, let $R_{i.}^2$ be the squared multiple correlation of X_i with the other X variables. Then

$$R_{i.}^2 = 1 - \frac{1}{c_{ii}} \tag{12.8}$$

where c_{ii} is the diagonal element of $\mathbf{R}^{-1}$ corresponding to $R_{i.}^2$. For example, if we want $R_{1.}^2$, we have $c_{11} = 1.26841$ and

$$R_{1.23}^2 = 1 - \frac{1}{1.26841} = .2116$$

Note that a very large diagonal element in $\mathbf{R}^{-1}$ is indicative that the variable has a very large squared multiple correlation with the remaining X variables.

12.7 Standard Errors of the Standardized Regression Coefficients

If all variables are in standardized form, then the standard error of a standardized regression coefficient will be given by

$$s_{\tilde{b}_i} = \sqrt{\frac{1 - R_{YY'}^2}{(n - k - 1)(1 - R_{i.}^2)}} \tag{12.9}$$

For X_1 we have $R_{i.}^2 = .2116$. We also have $R_{YY'}^2 = .611$. Then for the standard error of $\tilde{b}_1$, we have

$$s_{\tilde{b}_1} = \sqrt{\frac{1 - .611}{(10 - 3 - 1)(1 - .2116)}} = .28676$$

The standard error of $\tilde{b}_i$ will also be given by

$$s_{\tilde{b}_i} = \sqrt{\frac{c_{ii}(1 - R_{YY'}^2)}{n - k - 1}} \tag{12.10}$$

where c_{ii} is the diagonal element of $\mathbf{R}^{-1}$ corresponding to $\tilde{b}_i$. For $\tilde{b}_1$ we have $c_{11} = 1.26841$. Then substituting in (12.10) we have

$$s_{\tilde{b}_1} = \sqrt{\frac{(1.26841)(1 - .611)}{6}} = .28677$$

which is equal, within rounding errors, to the value we obtained with (12.9). For a test of significance of $\tilde{b}_1 = .54515$, we then obtain

$$t = \frac{.54515}{.28676} = 1.901$$

which is equal to the t test of the unstandardized coefficient, as shown in Table 11.2.

Exercises

12.1 Prove that if $z = (X - \bar{X})/s$, then $\bar{z} = 0$ and $s_z^2 = 1$.

12.2 Explain why if $z_1 = (X_1 - \bar{X}_1)/s_1$ and $z_2 = (X_2 - \bar{X}_2)/s_2$, then $c_{z_1 z_2} = r_{12}$.

12.3 Prove that if we have only one X variable, then $\tilde{b} = r_{YX}$.

12.4 Prove that with only one X variable, then $\Sigma y'^2/(n - 1) = r_{YX}^2$.

12.5 Table 9.1 in Chapter 9 has all the necessary data for calculating r_{Y1}, r_{Y2}, and r_{12}.

(a) Calculate these values. With only two X variables we have

$$\mathbf{R} = \begin{bmatrix} 1 & r_{12} \\ r_{12} & 1 \end{bmatrix}$$

(b) Calculate the inverse of $\mathbf{R}$, using the method shown for a 2×2 matrix in Chapter 10.

(c) Calculate the values of $\tilde{b}_1$ and $\tilde{b}_2$.

(d) Use the standardized values, $\tilde{b}_1$ and $\tilde{b}_2$, and formula (12.2) to obtain the unstandardized regression coefficients. They should be equal, within rounding errors, to the values given in Chapter 9 for the unstandardized regression coefficients, that is, $b_1 = 2.2857$ and $b_2 = .8571$.

12.6 We have $n = 10$ values of Y and $k = 3$ X variables as shown below.

Y	X_1	X_2	X_3
4	1	5	9
3	2	7	4
5	2	9	7
8	3	11	8
10	3	7	7
7	4	5	5
12	4	9	9
11	5	10	12
12	6	11	13
14	6	14	11

The variables were standardized and the intercorrelations of the X variables are as shown below.

$$\mathbf{R} = \begin{bmatrix} 1.00000 & .68512 & .66756 \\ .68512 & 1.00000 & .63969 \\ .66756 & .63969 & 1.00000 \end{bmatrix}$$

The inverse of $\mathbf{R}$ can be shown to be

$$\mathbf{R}^{-1} = \begin{bmatrix} 2.26438 & -.98919 & -.87882 \\ -.98919 & 2.12477 & -.69886 \\ -.87882 & -.69886 & 2.03372 \end{bmatrix}$$

For the same data we also have: $r_{Y1} = .883$, $r_{Y2} = .702$, and $r_{Y3} = .716$.
(a) Show that $\mathbf{R}^{-1}\mathbf{R} = \mathbf{I}$, within rounding errors.
(b) Calculate the standardized regression coefficients.
(c) Calculate the regression sum of squares.
(d) Calculate $R_{YY'}^2$.
(e) Test $R_{YY'}^2$ for significance.
(f) Calculate the standard error of each of the regression coefficients and test each one for significance.
(g) What is the value of the squared multiple correlation of X_1 with X_2 and X_3?

As you may have observed, the values of Y and X in this exercise are the same as those in Exercise 11.1 in the preceding chapter. The following questions ask you to compare certain results obtained in the two exercises.

(h) Are the two values of $R_{YY'}^2$ equal?
(i) Do the t tests of the standardized regression coefficients give the same results as the t tests of the unstandardized regression coefficients?
(j) Is the value of R_1^2 the same in both analyses?

Thirteen

Multiple Regression with Dummy and Effect Coding

13.1 Introduction

In this chapter we describe a multiple regression analysis involving a dependent variable Y and a nominal or categorical variable with k classes. The k classes can be k treatments in an experiment, k groups of individuals classified according to religious preference, or any other set of k mutually exclusive classes to which individuals belong or can be assigned. In analysis of variance terms, the k classes are commonly referred to as treatments and we shall also use this terminology. The total sum of squares (SS_{tot}) in the analysis of variance is partitioned into two components: the treatment sum of squares (SS_T) and the within treatment sum of squares (SS_W).

With a multiple regression analysis, the total sum of squares (SS_{tot}) is also partitioned into two components: the sum of squares for regression (SS_{reg}) and the residual sum of squares (SS_{res}). As we shall see, SS_{reg} is equal to SS_T and SS_{res} is equal to SS_W.

13.2 The Analysis of Variance for k Groups

Table 13.1 gives the $n_i = 5$ values of Y for each of $k = 4$ classes or treatments. In analyzing the data for this example, we first find the sum of squared deviations of the Y values from $\overline{Y}$, the overall mean of the Y values. This sum of squares, as we know, is simply $\Sigma(Y - \overline{Y})^2$ or SS_{tot} and will be given by

$$SS_{tot} = \Sigma(Y - \overline{Y})^2 = \Sigma Y^2 - \frac{(\Sigma Y)^2}{n} \qquad (13.1)$$

where $n = n_1 + n_2 + \cdots + n_k$.

For the data in Table 13.1 we have

$$SS_{tot} = (8)^2 + (7)^2 + \cdots + (2)^2 - \frac{(100)^2}{20} = 82.0$$

with $n - 1 = 20 - 1 = 19$ d.f.

TABLE 13.1   Values of Y for Each of $k = 4$ Groups

	Groups			
	1	2	3	4
	8	8	7	5
	7	7	5	4
	9	4	4	4
	6	4	3	3
	5	3	2	2
Σ	35	26	21	18

We then calculate the treatment sum of squares, or

$$SS_T = n_i \Sigma (\overline{Y}_i - \overline{Y})^2 = \frac{(\Sigma Y_1)^2}{n_1} + \frac{(\Sigma Y_2)^2}{n_2} + \cdots + \frac{(\Sigma Y_k)^2}{n_k} - \frac{(\Sigma Y)^2}{n}$$

(13.2)

where $n = n_1 + n_2 + \cdots + n_k$.

In our example, we have

$$SS_T = \frac{(35)^2}{5} + \frac{(26)^2}{5} + \frac{(21)^2}{5} + \frac{(18)^2}{5} - \frac{(100)^2}{20} = 33.2$$

with $k - 1 = 4 - 1 = 3$ d.f.

For each of the k classes or treatments we can find the sum of squared deviations of the n_i values of Y from the mean value, $\overline{Y}_i$, associated with that class. We are primarily interested in the sum of these sums of squared deviations. This pooled sum of squares can be obtained by subtraction and is commonly called the within treatment or within class sum of squares (SS_W). Thus

$$SS_W = SS_{tot} - SS_T$$

(13.3)

and SS_W will have $\Sigma n_i - k$ d.f. With an equal number of observations in each class, the degrees of freedom will simply be $kn_i - k = k(n_i - 1)$. In our example, we have

$$SS_W = 82.0 - 33.2 = 48.8$$

with $4(5 - 1) = 16$ d.f. If we divide SS_W by its degrees of freedom we obtain an unbiased estimate of the population variance σ_Y^2 of the Y values associated with each of the k classes or treatments.[1] This variance estimate is a mean square and is commonly referred to as *mean square within treatments* or classes. Thus

$$MS_W = \frac{SS_W}{n - k}$$

(13.4)

where again $n = n_1 + n_2 + \cdots + n_k$.

In our example, we have

$$MS_W = \frac{48.8}{16} = 3.05$$

If the Y means associated with each of the k treatments or classes are in no way dependent on the particular class, that is, if $\mu_1 = \mu_2 = \cdots = \mu_k$, then it can be shown that the *treatment mean square* or

$$MS_T = \frac{SS_T}{k - 1}$$

(13.5)

[1] We assume that the population variance σ_Y^2 is the same for each of the k classes.

is also an unbiased estimate of the same population variance as that estimated by MS_W. In our example, we have

$$MS_T = \frac{33.2}{4-1} = 11.07$$

and MS_T is considerably larger than MS_W.

To determine whether MS_T is significantly larger than MS_W, we have

$$F = \frac{MS_T}{MS_W} \tag{13.6}$$

with $k - 1$ d.f. for the numerator and $n - k$ d.f. for the denominator. In our example, we have

$$F = \frac{11.07}{3.05} = 3.63$$

a significant value with 3 and 16 d.f. and with $\alpha = .05$. It is reasonable to conclude that the Y means are not independent of the k classes or treatments.

13.3 Multiple Regression Analysis for k Groups with Dummy Coding

Table 13.2 repeats the Y values shown in Table 13.1. Table 13.2 also shows the unit column vector $\mathbf{X}_0$ and $k = 3$ $\mathbf{X}$ vectors. If we have k categories or classes, we can always represent the k classes by means of $k - 1$ $\mathbf{X}$ vectors. There are many ways in which these $\mathbf{X}$ vectors can be formed. Table 13.2 illustrates one commonly used method called *dummy coding* or 1, 0 coding.

In $\mathbf{X}_1$ each observation is assigned 1 if it belongs in class 1, and 0 otherwise. In $\mathbf{X}_2$ each observation is assigned 1 if it belongs in class 2, and 0 otherwise. In $\mathbf{X}_3$ each observation is assigned 1 if it belongs in class 3, and 0 otherwise. Note that all observations in the kth or last class have 0s in all three $\mathbf{X}$ vectors.

Consider now what the 4×4 symmetric $\mathbf{X'X}$ matrix will look like. Note that $\Sigma X_1 X_2 = \Sigma X_1 X_3 = \Sigma X_2 X_3 = 0$ and that $\Sigma X_1^2 = n_1$, $\Sigma X_2^2 = n_2$, and $\Sigma X_3^2 = n_3$. We also have $\Sigma X_1 = n_1$, $\Sigma X_2 = n_2$, and $\Sigma X_3 = n_3$. Then, we have

$$\mathbf{X'X} = \begin{bmatrix} n & \Sigma X_1 & \Sigma X_2 & \Sigma X_3 \\ \Sigma X_1 & \Sigma X_1^2 & \Sigma X_1 X_2 & \Sigma X_1 X_3 \\ \Sigma X_2 & \Sigma X_2 X_1 & \Sigma X_2^2 & \Sigma X_2 X_3 \\ \Sigma X_3 & \Sigma X_3 X_1 & \Sigma X_3 X_2 & \Sigma X_3^2 \end{bmatrix} = \begin{bmatrix} n & n_1 & n_2 & n_3 \\ n_1 & n_1 & 0 & 0 \\ n_2 & 0 & n_2 & 0 \\ n_3 & 0 & 0 & n_3 \end{bmatrix}$$

where $n = n_1 + n_2 + n_3 + n_4$. Then, for our example, we have

$$\mathbf{X'X} = \begin{bmatrix} 20 & 5 & 5 & 5 \\ 5 & 5 & 0 & 0 \\ 5 & 0 & 5 & 0 \\ 5 & 0 & 0 & 5 \end{bmatrix}$$

It can be shown that the inverse of $\mathbf{X'X}$ is equal to

$$(\mathbf{X'X})^{-1} = \begin{bmatrix} .2 & -.2 & -.2 & -.2 \\ -.2 & .4 & .2 & .2 \\ -.2 & .2 & .4 & .2 \\ -.2 & .2 & .2 & .4 \end{bmatrix}$$

TABLE 13.2   Values of Y, the Unit Vector X_0, and $k = 3$ X Vectors with Dummy Coding

	Y	X_0	X_1	X_2	X_3
	8	1	1	0	0
	7	1	1	0	0
Group 1	9	1	1	0	0
	6	1	1	0	0
	5	1	1	0	0
	8	1	0	1	0
	7	1	0	1	0
Group 2	4	1	0	1	0
	4	1	0	1	0
	3	1	0	1	0
	7	1	0	0	1
	5	1	0	0	1
Group 3	4	1	0	0	1
	3	1	0	0	1
	2	1	0	0	1
	5	1	0	0	0
	5	1	0	0	0
Group 4	4	1	0	0	0
	3	1	0	0	0
	2	1	0	0	0

Note also that

$$\mathbf{X'Y} = \begin{bmatrix} \Sigma Y \\ \Sigma X_1 Y \\ \Sigma X_2 Y \\ \Sigma X_3 Y \end{bmatrix} = \begin{bmatrix} \Sigma Y \\ \Sigma Y_1 \\ \Sigma Y_2 \\ \Sigma Y_3 \end{bmatrix} = \begin{bmatrix} 100 \\ 35 \\ 26 \\ 21 \end{bmatrix}$$

Solving for the least squares estimates, a, b_1, b_2, and b_3, we have $(\mathbf{X'X})^{-1}\mathbf{X'Y} = \mathbf{b}$ or

$$a = (.2)(100) + (-.2)(35) + (-.2)(26) + (-.2)(21) = 3.6$$

$$b_1 = (-.2)(100) + (.4)(35) + (.2)(26) + (.2)(21) \quad = 3.4$$

$$b_2 = (-.2)(100) + (.2)(35) + (.4)(26) + (.2)(21) \quad = 1.6$$

$$b_3 = (-.2)(100) + (.2)(35) + (.2)(26) + (.4)(21) \quad = .6$$

Observe now that $a = \overline{Y}_k = 3.6$, $b_1 = \overline{Y}_1 - \overline{Y}_k = 7.0 - 3.6 = 3.4$, $b_2 = \overline{Y}_2 - \overline{Y}_k = 5.2 - 3.6 = 1.6$, and $b_3 = \overline{Y}_3 - \overline{Y}_k = 4.2 - 3.6 = .6$.

The results shown above are perfectly general. It will always be true, with dummy coding, that a will be equal to the mean of the group which has 0 in each of the $\mathbf{X}$ vectors or the kth group; b_1 will always be equal to $\overline{Y}_1 - \overline{Y}_k$, b_2 will always be equal to $\overline{Y}_2 - \overline{Y}_k$, and so on. Nor do the results depend on having equal n_i's for the k groups. They apply also to the case where the n_i's are unequal. In general, then, with dummy coding, we have

$$a = \overline{Y}_k \tag{13.7}$$

and

$$b_i = \overline{Y}_i - \overline{Y}_k \tag{13.8}$$

where $\overline{Y}_k$ is the group with 0s in all $\mathbf{X}$ vectors.

For the regression equation, we have

$$Y' = a + b_1 X_1 + b_2 X_2 + b_3 X_3$$

or, in our example,

$$Y' = \overline{Y}_k + (\overline{Y}_1 - \overline{Y}_k)X_1 + (\overline{Y}_2 - \overline{Y}_k)X_2 + (\overline{Y}_3 - \overline{Y}_k)X_3$$

For each of the n_1 observations in group 1, we have $X_1 = 1$ and $X_2 = X_3 = 0$. Then, it is obvious that for these n_1 observations, $Y' = \overline{Y}_1$. Similarly, the n_2 observations in group 2 have $X_2 = 1$ and $X_1 = X_3 = 0$. Consequently, for each of these n_2 observations, $Y' = \overline{Y}_2$. Similarly, for the n_3 observations in group 3, $Y' = \overline{Y}_3$. For the n_k observations in the kth or last group, $X_1 = X_2 = X_3 = 0$, and consequently, for the observations in this group, $Y' = a = \overline{Y}_k$.

Because $Y' = \overline{Y}_i$ for each group, we have for each group,

$$\Sigma (Y - Y')^2 = \Sigma (Y - \overline{Y}_i)^2$$

or just the sum of squared deviations from the group mean. Summing these sums of squares over all k groups, we have

$$SS_{res} = SS_W$$

and, in our example, $SS_{res} = 48.8$. We have $SS_{tot} = 82.0$ and consequently,

$$SS_{reg} = SS_{tot} - SS_{res} = SS_T$$

or, in our example,

$$SS_{reg} = 82.0 - 48.8 = 33.2$$

We could also calculate

$$SS_{reg} = \mathbf{b'X'Y} - \frac{(\Sigma Y)^2}{n}$$

In our example, we have

$$\mathbf{b'X'Y} = (3.6)(100) + (3.4)(35) + (1.6)(26) + (.6)(21) = 533.2$$

We have $\Sigma Y = 100$ and $n = 20$. Then,

$$SS_{reg} = 533.2 - \frac{(100)^2}{20} = 33.2$$

as before.

For the squared multiple correlation, we obtain

$$R^2_{YY'} = \frac{SS_{reg}}{SS_{tot}} = \frac{33.2}{82.0} = .4049$$

and we can say that approximately 40 percent of the total sum of squares can be accounted for by the k classes or treatments. For the test of significance of $R^2_{YY'}$, we have

$$F = \frac{R^2_{Y.123}/k}{(1 - R^2_{Y.123})/(n - k - 1)}$$

or

$$F = \frac{.4049/3}{(1 - .4049)/(20 - 3 - 1)} = 3.63$$

which, of course, is equal to $F = MS_T/MS_W$.

13.4 Multiple Regression Analysis for *k* Groups with Effect Coding

Table 13.3 repeats the Y values and the unit column vector $\mathbf{X}_0$ shown in Table 13.2, but with $k = 3$ new $\mathbf{X}$ vectors. The $\mathbf{X}$ vectors in Table 13.3 were formed

TABLE 13.3   Values of Y, the Unit Vector X_0, and $k = 3$ X Vectors with Effect Coding

	Y	X_0	X_1	X_2	X_3
	8	1	1	0	0
	7	1	1	0	0
Group 1	9	1	1	0	0
	6	1	1	0	0
	5	1	1	0	0
	8	1	0	1	0
	7	1	0	1	0
Group 2	4	1	0	1	0
	4	1	0	1	0
	3	1	0	1	0
	7	1	0	0	1
	5	1	0	0	1
Group 3	4	1	0	0	1
	3	1	0	0	1
	2	1	0	0	1
	5	1	-1	-1	-1
	4	1	-1	-1	-1
Group 4	4	1	-1	-1	-1
	3	1	-1	-1	-1
	2	1	-1	-1	-1

using what is often called *effect coding* or $1, 0, -1$ coding. In effect coding we assign -1 to the observations in the kth or last group on all **X** vectors. For **X'X**, we have

$$\mathbf{X'X} = \begin{bmatrix} n & \Sigma X_1 & \Sigma X_2 & \Sigma X_3 \\ \Sigma X_1 & \Sigma X_1^2 & \Sigma X_1 X_2 & \Sigma X_1 X_3 \\ \Sigma X_2 & \Sigma X_2 X_1 & \Sigma X_2^2 & \Sigma X_2 X_3 \\ \Sigma X_3 & \Sigma X_3 X_1 & \Sigma X_3 X_2 & \Sigma X_3^2 \end{bmatrix}$$

where $n = n_1 + n_2 + \cdots + n_k$. If n_i is the same for each group, then it is easy to see in Table 13.3 that $\Sigma X_1 = \Sigma X_2 = \Sigma X_3 = 0$ and that $\Sigma X_1^2 = \Sigma X_2^2 = \Sigma X_3^2 = 2n_i$. Then, for the data in Table 13.3, we have

$$\mathbf{X'X} = \begin{bmatrix} 20 & 0 & 0 & 0 \\ 0 & 10 & 5 & 5 \\ 0 & 5 & 10 & 5 \\ 0 & 5 & 5 & 10 \end{bmatrix}$$

The inverse of $\mathbf{X'X}$ can be shown to be

$$(\mathbf{X'X})^{-1} = \begin{bmatrix} .05 & .00 & .00 & .00 \\ .00 & .15 & -.05 & -.05 \\ .00 & -.05 & .15 & -.05 \\ .00 & -.05 & -.05 & .15 \end{bmatrix}$$

We know that

$$\Sigma xy = \Sigma(X - \bar{X})(Y - \bar{Y})$$

$$= \Sigma XY - \frac{(\Sigma X)(\Sigma Y)}{n}$$

and it is obvious that if $\Sigma X = 0$, then $\Sigma xy = \Sigma XY$. For each of the $k = 3$ $\mathbf{X}$ vectors in Table 13.3, we have $\Sigma X_i = 0$. Consequently,

$$\mathbf{X'Y} = \begin{bmatrix} \Sigma Y \\ \Sigma X_1 Y \\ \Sigma X_2 Y \\ \Sigma X_3 Y \end{bmatrix} = \begin{bmatrix} \Sigma Y \\ \Sigma x_1 y \\ \Sigma x_2 y \\ \Sigma x_3 y \end{bmatrix} = \begin{bmatrix} \Sigma Y \\ \Sigma Y_1 - \Sigma Y_4 \\ \Sigma Y_2 - \Sigma Y_4 \\ \Sigma Y_3 - \Sigma Y_4 \end{bmatrix} = \begin{bmatrix} 100 \\ 17 \\ 8 \\ 3 \end{bmatrix}$$

Solving for the least squares estimates, a, b_1, b_2, and b_3, we have $(\mathbf{X'X})^{-1}\mathbf{X'Y} = \mathbf{b}$ or

$$a = (.05)(100) + (.00)(17) + (.00)(8) + (.00)(3) = 5.0$$

$$b_1 = (.00)(100) + (.15)(17) + (-.05)(8) + (-.05)(3) = 2.0$$

$$b_2 = (.00)(100) + (-.05)(17) + (.15)(8) + (-.05)(3) = .2$$

$$b_3 = (.00)(100) + (-.05)(17) + (-.05)(8) + (.15)(3) = -.8$$

Note now that $a = \bar{Y} = 5.0$, $b_1 = \bar{Y}_1 - \bar{Y} = 7.0 - 5.0 = 2.0$, $b_2 = \bar{Y}_2 - \bar{Y} = 5.2 - 5.0 = .2$, and $b_3 - \bar{Y}_3 - \bar{Y} = 4.2 - 5.0 = -.8$. For the regression equation, we have

$$Y' = a + b_1 X_1 + b_2 X_2 + b_3 X_3$$

or

$$Y' = \bar{Y} + (\bar{Y}_1 - \bar{Y})X_1 + (\bar{Y}_2 - \bar{Y})X_2 + (\bar{Y}_3 - \bar{Y})X_3$$

For the n_1 observations in the first group, this reduces to

$$Y' = \bar{Y} + (\bar{Y}_1 - \bar{Y})X_1 = \bar{Y}_1 = 7.0$$

For the n_2 observations in the second group, the regression equation reduces to

$$Y' = \bar{Y} + (\bar{Y}_2 - \bar{Y})X_2 = \bar{Y}_2 = 5.2$$

Similarly, for the n_3 observations in the third group, we have

$$Y' = \overline{Y} + (\overline{Y}_3 - \overline{Y})X_3 = \overline{Y}_3 = 4.2$$

For the n_k observations in the kth or last group, we see that $X_1 = X_2 = X_3 = -1$, so that for this group

$$Y' = \overline{Y} - (\overline{Y}_1 - \overline{Y}) - (\overline{Y}_2 - \overline{Y}) - (\overline{Y}_3 - \overline{Y})$$

$$= 4\overline{Y} - (\overline{Y}_1 + \overline{Y}_2 + \overline{Y}_3)$$

$$= \frac{4(\overline{Y}_1 + \overline{Y}_2 + \overline{Y}_3 + \overline{Y}_4) - 4(\overline{Y}_1 + \overline{Y}_2 + \overline{Y}_3)}{4}$$

$$= \overline{Y}_4$$

Because $Y' = \overline{Y}_i$ for the n_i observations in each group, we also have

$$SS_{res} = \Sigma(Y - Y')^2 = SS_W = 48.8$$

and, because $SS_{tot} - SS_{res} = SS_{reg}$, we also have

$$SS_{reg} = \Sigma(Y' - \overline{Y})^2 = SS_T = 33.2$$

We also know that

$$SS_{reg} = b_1\Sigma x_1 y + b_2\Sigma x_2 y + \cdots + b_k\Sigma x_k y$$

and consequently, in our example, we have

$$SS_{reg} = (2.0)(17) + (.2)(8) + (-.8)(3) = 33.2$$

as before.

Note that we obtained exactly the same values for SS_{reg} and SS_{res} using effect coding as we did when we used dummy coding. Consequently, no matter which method of coding we use, we will obtain exactly the same value for $R^2_{YY'}$ and for the F test of significance of $R^2_{YY'}$.

13.4 Effect Coding with Unequal n_i's

With effect coding and *unequal* n_i's for the k groups or classes, a will be equal to the *unweighted* mean of the k means or

$$a = \frac{\overline{Y}_1 + \overline{Y}_2 + \cdots + \overline{Y}_k}{k} \tag{13.8}$$

The values of the regression coefficients will then be

$$b_i = \overline{Y}_i - a \tag{13.9}$$

where a is defined by (13.8). Only with equal n_i's for the k groups will the unweighted mean defined by (13.8) be equal to the overall or weighted mean

$$\overline{Y} = \frac{n_1\overline{Y}_1 + n_2\overline{Y}_2 + \cdots + n_k\overline{Y}_k}{n_1 + n_2 + \cdots + n_k} \qquad (13.10)$$

Thus, with equal n_i's, we have

$$b_i = \overline{Y}_i - a = \overline{Y}_i - \overline{Y}$$

Exercises

13.1   We have $n_i = 5$ values of Y for each of $k = 3$ classes or groups as shown below:

	Groups	
1	2	3
12	9	6
11	9	7
10	7	3
10	6	2
7	4	2

(a) Calculate SS_T and SS_W and test MS_T for significance.
(b) Use a unit vector $\mathbf{X}_0$ and two vectors with dummy coding. Calculate $\mathbf{X'X}$.
(c) The inverse of $\mathbf{X'X}$ can be shown to be

$$(\mathbf{X'X})^{-1} = \begin{bmatrix} .2 & -.2 & -.2 \\ -.2 & .4 & .2 \\ -.2 & .2 & .4 \end{bmatrix}$$

Calculate $(\mathbf{X'X})^{-1}\mathbf{X'X} = \mathbf{I}$.
(d) Calculate $\mathbf{X'Y}$.
(e) Calculate $(\mathbf{X'X})^{-1}\mathbf{X'Y} = \mathbf{b}$. You should find that $a = \overline{Y}_3$, $b_1 = \overline{Y}_1 - \overline{Y}_3$, and $b_2 = \overline{Y}_2 - \overline{Y}_3$.
(f) Write the regression equation using the values in $\mathbf{b}$.
(g) Calculate SS_{reg} and SS_{res}.
(h) Calculate $R_{YY'}^2$ and test $R_{YY'}^2$ for significance. Is the obtained value of F equal to the F test of MS_T?
(i) Explain why SS_{res} is equal to SS_W.

13.2   For the data in Exercise 13.1 use a unit vector $\mathbf{X}_0$ and two vectors with effect coding.
(a) Calculate $\mathbf{X'X}$.

(b) The inverse of $X'X$ can be shown to be

$$(X'X)^{-1} = \begin{bmatrix} 1/15 & 0 & 0 \\ 0 & 2/15 & -1/15 \\ 0 & -1/15 & 2/15 \end{bmatrix}$$

Calculate $(X'X)^{-1}(X'X) = I$.

(c) Calculate $X'Y$.

(d) Calculate $(X'X)^{-1}X'Y = b$. You should find that $a = \overline{Y}$, $b_1 = \overline{Y}_1 - \overline{Y}$, and $b_2 = \overline{Y}_2 - \overline{Y}$.

(e) Write the regression equation using the values in b.

(f) Calculate SS_{reg} and SS_{res}.

(g) Calculate $R_{YY'}^2$ and test $R_{YY'}^2$ for significance. Is the obtained value of F equal to the F test of MS_T?

(h) Explain why SS_{res} is equal to SS_W.

13.3 We have $k = 4$ groups with unequal n_i's as shown below.

	Groups		
1	2	3	4
2	4	4	8
4	5	8	9
	6		10
			11
			12

(a) Calculate SS_W and SS_T and test MS_T for significance.

(b) Use a unit vector X_0 and three vectors with effect coding. Calculate $X'X$.

(c) The inverse of $X'X$ can be shown to be

$$(X'X)^{-1} = \begin{bmatrix} .095833 & .029167 & -.012500 & .029167 \\ .029167 & .345833 & -.112500 & -.154167 \\ -.012500 & -.112500 & .262500 & -.112500 \\ .029167 & -.154167 & -.112500 & .345833 \end{bmatrix}$$

Calculate $X'Y$ and $(X'X)^{-1}X'Y = b$.

(d) Write the regression equation using the values in b. You should find that $a = \overline{Y}$, $b_1 = \overline{Y}_1 - \overline{Y}$, $b_2 = \overline{Y}_2 - \overline{Y}$, and $b_3 = \overline{Y}_3 - \overline{Y}$, where $\overline{Y}$ is the *unweighted mean of the k means*.

(e) Calculates SS_{res} and SS_{reg}.

(f) Calculate $R_{YY'}^2$ and test $R_{YY'}^2$ for significance. Is the obtained value of F equal to the F test of MS_T?

13.4 For the data in Exercise 13.3 use a unit vector X_0 and three vectors with dummy coding.

(a) Calculate $X'X$.

(b) The inverse of $\mathbf{X'X}$ can be shown to be

$$(\mathbf{X'X})^{-1} = \begin{bmatrix} .200000 & -.200000 & -.200000 & -.200000 \\ -.200000 & .700000 & .200000 & .200000 \\ -.200000 & .200000 & .533333 & .200000 \\ -.200000 & .200000 & .200000 & .700000 \end{bmatrix}$$

Calculate $\mathbf{X'Y}$ and $(\mathbf{X'X})^{-1}\mathbf{X'Y} = \mathbf{b}$.

(c) Write the regression equation using the values in $\mathbf{b}$. You should find that $a = \overline{Y}_4$, $b_1 = \overline{Y}_1 - \overline{Y}_4$, $b_2 = \overline{Y}_2 - \overline{Y}_4$, and $b_3 = \overline{Y}_3 - \overline{Y}_4$.

(d) Calculates SS_{res} and SS_{reg}.

(e) Calculate $R_{YY'}^2$ and test $R_{YY'}^2$ for significance.

Fourteen

Tests for the Trend of a Set of k Sample Means

14.1 Introduction

Consider an experiment involving k equally spaced values of a quantitative variable X. For example, if X is a measure of time, the values of X might be 5 seconds, 10 seconds, 15 seconds, and 20 seconds, or they might be 2 minutes, 4 minutes, 6 minutes, and 8 minutes. In a learning experiment the values of X might consist of 1, 2, 3, 4, and 5 trials. If X is a dosage of a drug, then the values of X might be 1 grain, 2 grains, 3 grains, and so on. We assume that not only are the values of X equally spaced but they are also fixed; that is, if the experiment were to be repeated, the values of X would be the same.

For each value of X we obtain n_i independent random observations of a dependent variable Y. The values of Y are assumed to be normally distributed for each value of X with the same variance $\sigma^2_{Y \cdot X} = \sigma^2_Y$. The sample Y means obtained for each value of X are unbiased estimates of the corresponding population means. On the basis of the sample means we wish to infer something about the trend of the population means over the fixed values of X.

We let the ordered sums of the Y values be represented by $\Sigma Y_1, \Sigma Y_2, \ldots,$ ΣY_k and the ordered means be represented by $\overline{Y}_1, \overline{Y}_2, \ldots, \overline{Y}_k$. The sum of all the Y values will be represented by ΣY and the mean will be represented by $\overline{Y}$. Then the weighted sum of squared deviations of the means $\overline{Y}_i$ from the overall mean $\overline{Y}$ will be given by

$$SS_T = \sum_1^k n_i (\overline{Y}_i - \overline{Y})^2 \tag{14.1}$$

where SS_T is the treatment sum of squares with $k - 1$ d.f. As we know, an algebraic equivalent of (14.1) that may also simplify the calculations is

$$SS_T = \frac{(\Sigma Y_1)^2}{n_i} + \frac{(\Sigma Y_2)^2}{n_i} + \cdots + \frac{(\Sigma Y_k)^2}{n_i} - \frac{(\Sigma Y)^2}{kn_i} \tag{14.2}$$

If the Y means all fall on a straight line, with slope $b_i \neq 0$, then all of the variation in the Y means can be accounted for by an equation of the first degree or a linear equation. That part of SS_T that can be accounted for by an equation of the first degree is referred to as the *linear component* of SS_T or as SS_L with 1 d.f. Then the residual sum of squares or the sum of squares for deviations from linearity will be given by

$$SS_{res} = SS_T - SS_L$$

with $k - 2$ d.f.

If SS_{res} represents significant deviations of the Y means from linearity, then we may determine how much of this residual variation can be accounted for by an equation of the second degree or by the *quadratic component* of SS_T. Any set of k means can always be described accurately by an equation of degree no higher than $k - 1$. With $k = 5$ means, for example, it is always possible to partition SS_T into a linear, quadratic, cubic, and quartic component corresponding to an equation of the first, second, third, and fourth degree. Not all of these components will necessarily be required to describe the trend of the means. Some components may be equal to zero or sufficiently small as to be nonsignificant.

In most behavioral science experiments, we would ordinarily be interested in first determining whether the linear component is significant, and then in determining whether the deviations from linearity are significant. If they are, then we would determine whether the quadratic component (SS_Q) is significant, and if so, whether there are significant deviations from the linear and quadratic components, that is, whether the new residual sum of squares

$$SS_{res} = SS_T - SS_L - SS_Q$$

is significant. Occasionally, we may also be interested in the cubic component and the quartic component.

14.2   Coefficients for Orthogonal Polynomials

Instead of working with polynomial equations of the second degree and higher, we make use of coefficients for orthogonal polynomials. These coefficients enable us to investigate systematically the trend of the means and to find out which components of the trend are significant using only an equation of the first degree. Table VIII in the Appendix[1] gives the coefficients for orthogonal polynomials for $k = 3$ to $k = 10$ values of X. Table 14.1 gives these coefficients for $k = 5$ values of X. Note that there are only $k - 1 = 4$ sets of orthogonal coefficients corresponding to the linear, quadratic, cubic, and quartic components. These coefficients are represented by x because, for each

TABLE 14.1   Orthogonal Coefficients for the Linear, Quadratic, Cubic, and Quartic Components of the Trend for $k = 5$ Means

Component		Coefficients				
Linear:	x_1	−2	−1	0	1	2
Quadratic:	x_2	2	−1	−2	−1	2
Cubic:	x_3	−1	2	0	−2	1
Quartic:	x_4	1	−4	6	−4	1

[1]Table VIII gives only the coefficients for the linear, quadratic, cubic, and quartic components. Orthogonal coefficients for the higher-degree polynomials can be found in R. A. Fisher and F. Yates, *Statistical Tables for Biological, Agricultural and Medical Research* (3d ed.). Edinburgh: Oliver and Boyd, 1936.

set of coefficients, we have

$$\Sigma x_1 = \Sigma x_2 = \Sigma x_3 = \Sigma x_4 = 0$$

The sets of coefficients are described as *orthogonal* because, for any two sets (for example, x_1 and x_2), we have

$$\Sigma x_1 x_2 = 0$$

Thus the correlation between x_1 and x_2 will be equal to zero, and this is also true for any two sets, x_i and x_j, $i \neq j$.

We have said that any set of k means can be accurately described by an equation of degree no higher than $k - 1$. With coefficients for orthogonal polynomials this means that the linear equation

$$\overline{Y}_i' = \overline{Y} + b_1 x_1 + b_2 x_2 + \cdots + b_{k-1} x_{k-1} \qquad (14.3)$$

will result in a set of predicted values that are equal to the corresponding observed values, $\overline{Y}_i$, of the means. The successive terms $b_1 x_1$, $b_2 x_2$, $\ldots$, $b_{k-1} x_{k-1}$ correspond to terms of the first, second, third, and higher degree in a polynomial equation. Not all of the terms may be needed to describe accurately the trend of the means. For example, the trend of the means may be described accurately by

$$\overline{Y}_i' = \overline{Y} + b_1 x_1$$

or by an equation of the first degree. If both the linear and quadratic components of the trend are significant and the other components are not, then

$$\overline{Y}_i' = \overline{Y} + b_1 x_1 + b_2 x_2$$

may be assumed to describe the trend of the means. If only the quadratic component is significant, then

$$\overline{Y}_i' = \overline{Y} + b_2 x_2$$

may be assumed to describe the trend of the means.

In (14.3) the values of x_1, x_2, $\ldots$, x_{k-1} are known and can be obtained from Table VIII in the Appendix. The corresponding values of b_1, b_2, $\ldots$, b_{k-1} are unknown, but they can be obtained from the data of the experiment.

14.3 The Analysis of Variance for $k = 5$ Values of X

Table 14.2 gives the $n_i = 5$ values of Y for each of $k = 5$ equally spaced values of an independent variable X.[2] In analyzing the data for this example, we first

[2]The coefficients for orthogonal polynomials given in Table VIII assume that we have equally spaced values of X and that we have the same number of observations of Y for each value of X. If the values of X are not equally spaced, or if the n_i's are not equal, the coefficients given in Table

TABLE 14.2 Values of Y for $k = 5$ Equally Spaced Values of a Quantitative Variable X

	\multicolumn{5}{c}{X variable}
	1	2	3	4	5
	10	12	9	14	15
	8	14	13	13	13
	12	11	12	11	12
	11	10	10	12	16
	9	13	11	15	14
Σ	50	60	55	65	70

find the sum of squared deviations of the Y values from $\overline{Y}$, the overall mean of the Y values. This sum of squares, as we know, is called the total sum of squares and will be given by

$$SS_{tot} = \Sigma(Y - \overline{Y})^2 = \Sigma Y^2 - \frac{(\Sigma Y)^2}{kn_i} \qquad (14.4)$$

For the data in Table 14.2 we have

$$SS_{tot} = (10)^2 + (8)^2 + \cdots + (14)^2 - \frac{(300)^2}{(5)(5)} = 100$$

with $kn_i - 1$ d.f.

We then calculate the treatment sum of squares. In our example, we have

$$SS_T = \frac{(50)^2}{5} + \frac{(60)^2}{5} + \frac{(55)^2}{5} + \frac{(65)^2}{5} + \frac{(70)^2}{5} - \frac{(300)^2}{25}$$

$$= 3650 - 3600$$

$$= 50$$

and this sum of squares will have $k - 1$ d.f.

For each value of X we can now find the sum of squared deviations of the $n_i = 5$ values of Y from the mean value, $\overline{Y}_i$, associated with that value of X. We are primarily interested in the sum of these sums of squared deviations. This pooled sum of squares can be obtained by subtraction, as we have pointed out

---

VIII are not applicable. Coefficients for orthogonal polynomials can be derived when the values of the independent variable are not equally spaced or when the n_i's are not equal, but it would obviously be impractical to table these coefficients for all possible cases. See J. Gaito, Unequal intervals and unequal n's in trend analyses, *Psychological Bulletin*, 1965, *63*, 125–127, for methods that can be used to derive the coefficients for unequal n_i's or for values of X that are not equally spaced.

previously, and is commonly referred to as the within treatment sum of squares, or SS_W. Thus

$$SS_W = SS_{tot} - SS_T \tag{14.5}$$

and SS_W will have $k(n_i - 1)$ d.f. In our example, we have

$$SS_W = 100 - 50 = 50$$

with $5(5 - 1) = 20$ d.f. If we divide SS_W by its degrees of freedom, we obtain an unbiased estimate of the common population variance σ_Y^2 of the Y values associated with each value of X. This variance estimate is a mean square and is commonly referred to as the *mean square within treatments*. Thus

$$MS_W = \frac{SS_W}{k(n_i - 1)}$$

In our example, we have

$$MS_W = \frac{50}{5(5 - 1)} = 2.5$$

If the Y means associated with each value of X are in no way dependent on the particular values of X, that is, if $\mu_1 = \mu_2 = \cdots = \mu_k$, then it can be shown that the *treatment mean square* or

$$MS_T = \frac{SS_T}{k - 1}$$

is also an unbiased estimate of the same population variance as that estimated by MS_W. In our example, we have

$$MS_T = \frac{50}{5 - 1} = 12.5$$

and MS_T is considerably larger than MS_W.

To determine whether MS_T is significantly larger than MS_W, we have

$$F = \frac{MS_T}{MS_W}$$

with $k - 1$ d.f. for the numerator and $k(n_i - 1)$ d.f. for the denominator. In our example, we have

$$F = \frac{12.5}{2.5} = 5.0$$

a significant value with $\alpha = .01$ and with 4 and 20 d.f. It is reasonable to conclude that the Y means are not independent of the values of X.

4.4 Components of the Trend

Our primary interest, however, is the trend of the means. Figure 14.1 shows a plot of the Y means against the values of X. We know that an equation of the form

$$\overline{Y}_i' = \overline{Y} + b_1 x_1 + b_2 x_2 + b_3 x_3 + b_4 x_4$$

where the values of x are the coefficients for orthogonal polynomials will predict perfectly the corresponding values of $\overline{Y}_i$. But some of the components of the trend represented by

$$\frac{[\Sigma(x_i \Sigma Y_i)]^2}{n_i \Sigma x_i^2}$$

may simply represent random variation and, in this instance, we may regard the values of

$$b_i = \frac{\Sigma(x_i \Sigma Y_i)}{n_i \Sigma x_i^2}$$

as not differing significantly from zero. Consequently, these terms may be dropped. We shall proceed systematically to investigate the components of the trend, starting with the linear component.

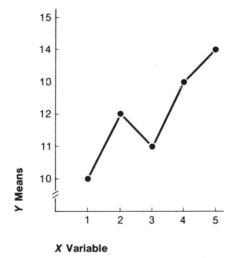

FIGURE 14.1   Plot of $k = 5$ means for equally spaced values of a quantitative variable X.

14.5 The Linear Component and a Test of Significance

To calculate the linear component of the trend, we use the coefficients x_1 for $k = 5$ as obtained from Table VIII in the Appendix. Then, we have

$$\frac{1}{n_i}\Sigma(x_1 \Sigma Y_i) = \frac{1}{5}[(-2)(50) + (-1)(60) + (0)(55) + (1)(65) + (2)(70)]$$

$$= \frac{1}{5}(45)$$

and

$$b_1 = \frac{\Sigma(x_1 \Sigma Y_i)}{n_i \Sigma x_1^2} = \frac{45}{(5)(10)} = .90$$

For the sum of squares for the linear component of the trend, we have

$$SS_L = \frac{[\Sigma(x_1 \Sigma Y_1)]^2}{n_i \Sigma x_1^2} = \frac{(45)^2}{(5)(10)} = 40.5$$

with 1 d.f. Because SS_L has only 1 d.f., the mean square for the linear component will be equal to SS_L, that is, $MS_L = SS_L/1 = SS_L$.[3] To determine whether the linear component is significant, we find

$$F = \frac{MS_L}{MS_W} = \frac{SS_L}{MS_W} = \frac{40.5}{2.5} = 16.2$$

with 1 and 20 d.f. Because $F = 16.2$ is highly significant, we conclude that there is a significant linear component in the trend of the means. Our regression equation will then be

$$\overline{Y}'_i = \overline{Y} + b_1 x_1$$

where $\overline{Y} = 12$ and $b_1 = .90$ and the values of x_1 are the coefficients for the linear component.

In our example, we have

$$\overline{Y}'_1 = 12 + (.90)(-2) = 10.20$$

$$\overline{Y}'_2 = 12 + (.90)(-1) = 11.10$$

$$\overline{Y}'_3 = 12 + (.90)(0) = 12.00$$

$$\overline{Y}'_4 = 12 + (.90)(1) = 12.90$$

$$\overline{Y}'_5 = 12 + (.90)(2) = 13.80$$

---

[3] In the analysis of variance, as we have pointed out previously, sums of squares divided by their degrees of freedom are commonly referred to as mean squares. When a sum of squares, such as SS_L, has only 1 d.f., the sum of squares is itself a mean square.

The residual sum of squares will then be given by

$$SS_{res} = n_i \Sigma (\overline{Y}_i - \overline{Y}'_i)^2 \qquad (14.6)$$

where $\overline{Y}'_i = \overline{Y} + b_1 x_1$. In our example, we have

$$SS_{res} = 5[(10 - 10.2)^2 + (12 - 11.1)^2 + (11 - 12.0)^2$$
$$+ (13 - 12.9)^2 + (14 - 13.8)^2]$$
$$= 9.5$$

Ordinarily we would not use (14.6) to calculate SS_{res} because this sum of squares can be obtained much more easily by subtraction.[4] Thus

$$SS_{res} = SS_T - SS_L \qquad (14.7)$$

In our example, we have $SS_T = 50.0$ and $SS_L = 40.5$ and, consequently,

$$SS_{res} = 50.0 - 40.5 = 9.5$$

which is equal to the value we obtained by direct calculation using (14.6).

SS_{res} measures the deviations of the Y means from linearity and will have $k - 2$ d.f. To determine whether $MS_{res} = SS_{res}/(k - 2)$ is significant, we find

$$F = \frac{MS_{res}}{MS_W}$$

and, in our example,

$$F = \frac{9.5/(5 - 2)}{2.5} = 1.27$$

with 3 and 20 d.f., and this is a nonsignificant value of F with $\alpha = .05$.

Because there is a significant linear component of the trend and a nonsignificant residual variance, we would customarily end the investigation at this point. For purposes of illustration, however, we proceed to calculate the quadratic, cubic, and quartic components of the trend.

14.6 The Quadratic Component and a Test of Significance

Using the coefficients x_2 for the quadratic component, we obtain

$$\frac{1}{n_i} \Sigma(x_2 \Sigma Y_i) = \frac{1}{5} [(2)(50) + (-1)(60) + (-2)(55) + (-1)(65) + (2)(70)]$$

$$= \frac{1}{5}(5)$$

---

[4]The proof of (14.7) is given in answer to one of the exercises at the end of this chapter.

Then

$$b_2 = \frac{\Sigma(x_2 \Sigma Y_i)}{n_i \Sigma x_2^2} = \frac{5}{(5)(14)} = .07$$

and

$$SS_Q = \frac{[\Sigma(x_2 \Sigma Y_i)]^2}{n_i \Sigma x_2^2} = \frac{(5)^2}{(5)(14)} = .36$$

with 1 d.f.

It is obvious that because

$$F = \frac{MS_Q}{MS_W} = \frac{.36}{2.50} < 1.00$$

the quadratic component is not significant. However, if we were to include the quadratic term in the regression equation, we would now have

$$\overline{Y}'_i = \overline{Y} + b_1 x_1 + b_2 x_2$$

or

$$\overline{Y}'_i = 12 + .90x_1 + .07x_2$$

and the resulting values of $\overline{Y}'_i$ would be

$$\overline{Y}'_1 = 10.34$$
$$\overline{Y}'_2 = 11.03$$
$$\overline{Y}'_3 = 11.86$$
$$\overline{Y}'_4 = 12.83$$
$$\overline{Y}'_5 = 13.94$$

The new residual sum of squares,

$$SS_{res} = n_i \Sigma (\overline{Y}_i - \overline{Y}'_i)^2$$

where $\overline{Y}'_i$ is now equal to $\overline{Y} + b_1 x_1 + b_2 x_2$, could be calculated directly and would be found to be equal to

$$SS_{res} = SS_T - SS_L - SS_Q$$

or

$$SS_{res} = 50.00 - 40.50 - .36 = 9.14$$

with $k - 3 = 2$ d.f.

It is obvious that taking the quadratic term into account in the regression equation has not reduced the residual sum of squares by any considerable amount.

14.7 The Cubic Component and a Test of Significance

Using the coefficients x_3 for the cubic component, we obtain

$$\frac{1}{n_i}\Sigma(x_3\Sigma Y_i) = \frac{1}{5}[(-1)(50) + (2)(60) + (0)(55) + (-2)(65) + (1)(70)]$$

$$= \frac{1}{5}(10)$$

Then

$$b_3 = \frac{\Sigma(x_3\Sigma Y_i)}{n_i\Sigma x_3^2} = \frac{10}{(5)(10)} = .20$$

and the sum of squares for the cubic component will be

$$SS_C = \frac{[\Sigma(x_3\Sigma Y_i)]^2}{n_i\Sigma x_3^2} = \frac{(10)^2}{(5)(10)} = 2.0$$

with 1 d.f.

A test of significance of the cubic component will be given by

$$F = \frac{MS_C}{MS_W}$$

In our example, $MS_C = 2.0$ and $MS_W = 2.5$; F is less than 1.00 and is nonsignificant.

Taking into account the linear, quadratic, and cubic terms, we have

$$\overline{Y}'_i = \overline{Y} + b_1x_1 + b_2x_2 + b_3x_3$$

or

$$\overline{Y}'_i = 12 + .90x_1 + .07x_2 + .20x_3$$

and the predicted means will now be

$$\overline{Y}'_1 = 10.14$$
$$\overline{Y}'_2 = 11.43$$
$$\overline{Y}'_3 = 11.86$$
$$\overline{Y}'_4 = 12.43$$
$$\overline{Y}'_5 = 14.14$$

If we were now to calculate the new residual sum of squares, using $\overline{Y}'_i = 12 + .90x_1 + .07x_2 + .20x_3$ as the predicted value of the Y means, we would find that

$$SS_{res} = n_i\Sigma(\overline{Y}_i - \overline{Y}'_i)^2 = SS_T - SS_L - SS_Q - SS_C$$

or, in our example,

$$SS_{res} = 50.00 - 40.50 - .36 - 2.00 = 7.14$$

with $k - 4 = 1$ d.f.

14.8 The Quartic Component and a Test of Significance

The sum of squares for the quartic component must be equal to 7.14 because we know that SS_T is equal to the sum of the linear, quadratic, cubic, and quartic components. To show that this is so, we use the coefficients x_4 for the quartic component to obtain

$$\frac{1}{n_i} \Sigma(x_4 \Sigma Y_i) = \frac{1}{5}[(1)(50) + (-4)(60) + (6)(55) + (-4)(65) + (1)(70)]$$

$$= \frac{1}{5}(-50)$$

and

$$b_4 = \frac{\Sigma(x_4 \Sigma Y_i)}{n_i \Sigma x_4^2} = \frac{-50}{(5)(70)} = -.14$$

Then the sum of squares for the quartic component will be

$$\frac{[\Sigma(x_4 \Sigma Y_i)]^2}{n_i \Sigma x_4^2} = \frac{(-50)^2}{(5)(70)} = 7.14$$

with 1 d.f.

As a test of significance of the quartic component, we have

$$F = \frac{7.14}{2.50} = 2.86$$

with 1 and 20 d.f., a nonsignificant value with $\alpha = .05$.

Taking into account the linear, quadratic, cubic, and quartic terms, we now have

$$\overline{Y}_i' = \overline{Y} + b_1 x_1 + b_2 x_2 + b_3 x_3 + b_4 x_4$$

or

$$\overline{Y}_i' = 12 + .90x_1 + .07x_2 + .20x_3 + .14x_4$$

and the predicted means are now

$$\overline{Y}_1' = 10.00$$
$$\overline{Y}_2' = 11.99$$
$$\overline{Y}_3' = 11.02$$
$$\overline{Y}_4' = 12.99$$
$$\overline{Y}_5' = 14.00$$

These means are equal, within rounding errors, to the observed values of the means.

14.9 Correlations of $\overline{Y}_i$ with the Orthogonal Coefficients

We now note that the sum of the squares of the correlation coefficients

$$r_{\overline{Y}_i x_1}^2 = \frac{[\Sigma(x_1 \Sigma Y_i)]^2}{(n_i \Sigma x_1^2)(SS_T)} = \frac{(45)^2}{(5)(10)(50)} = .810$$

$$r_{\overline{Y}_i x_2}^2 = \frac{[\Sigma(x_2 \Sigma Y_i)]^2}{(n_i \Sigma x_2^2)(SS_T)} = \frac{(-5)^2}{(5)(14)(50)} = .007$$

$$r_{\overline{Y}_i x_3}^2 = \frac{[\Sigma(x_3 \Sigma Y_i)]^2}{(n_i \Sigma x_3^2)(SS_T)} = \frac{(10)^2}{(5)(10)(50)} = .040$$

$$r_{\overline{Y}_i x_4}^2 = \frac{[\Sigma(x_4 \Sigma Y_i)]^2}{(n_i \Sigma x_4^2)(SS_T)} = \frac{(-50)^2}{(5)(70)(50)} = .143$$

is equal to 1.00. In this example, each of the sets of orthogonal coefficients accounts for a proportion of the treatment sum of squares (SS_T). The largest proportion (.81) is accounted for by the coefficients for the linear component.

4.10 Correlation Coefficient of $\overline{Y}_i$ with $\overline{Y}_i'$

The correlation coefficient between the actual Y means and the predicted values

$$\overline{Y}_i' = \overline{Y} + b_1 x_1 + b_2 x_2 + b_3 x_3 + b_4 x_4$$

is the multiple correlation coefficient between $\overline{Y}_i$ and a weighted linear sum of x_1, x_2, x_3, and x_4, the weights being the corresponding regression coefficients b_1, b_2, b_3, and b_4. Because the values of x_1, x_2, x_3, and x_4 are orthogonal or uncorrelated, the square of the multiple correlation coefficient will be

$$R_{\overline{Y}_i \overline{Y}_i'}^2 = r_{\overline{Y}_i x_1}^2 + r_{\overline{Y}_i x_2}^2 + r_{\overline{Y}_i x_3}^2 + r_{\overline{Y}_i x_4}^2$$

or, in our example,

$$R^2_{\bar{Y}_i \bar{Y}'_i} = .810 + .007 + .040 + .143 = 1.000$$

The square of the multiple correlation coefficient will always be equal to 1.00 if all $k - 1$ terms are used in the regression equation. If some of the b values are equal to zero so that the corresponding values of r are equal to zero, then $R^2_{\bar{Y}_i \bar{Y}'_i}$ will still be equal to 1.00, if all the remaining terms in the regression equation are used. It is, of course, possible for $R^2_{\bar{Y}_i \bar{Y}'_i}$ to be equal to 1.00 when only one term is included in the regression equation.

The square of the multiple correlation coefficient is simply the proportion of SS_T that can be accounted for by a weighted sum of the x values. For example, using only the x_1 and x_4 values, we have

$$Y' = \bar{Y} + b_1 x_1 + b_4 x_4$$

and

$$R^2_{\bar{Y}_i \bar{Y}'_i} = .810 + .143 = .953$$

Approximately 95 percent of the treatment sum of squares can be accounted for by the linear and quartic components.

14.11 Multiple Regression with Orthogonal Variables

Table 14.3 repeats the Y values given in Table 14.2 along with the unit column vector X_0 and four x vectors, x_1, x_2, x_3, and x_4, which contain the coefficients for the linear, quadratic, cubic, and quartic components of the trend. Note that $\Sigma x_1 = \Sigma x_2 = \Sigma x_3 = \Sigma x_4 = 0$ and that all vectors are mutually orthogonal, that is, $\Sigma x_i x_j = 0$. For $X'X$ we have

$$X'X = \begin{bmatrix} n & 0 & 0 & 0 & 0 \\ 0 & \Sigma x_1^2 & 0 & 0 & 0 \\ 0 & 0 & \Sigma x_2^2 & 0 & 0 \\ 0 & 0 & 0 & \Sigma x_3^2 & 0 \\ 0 & 0 & 0 & 0 & \Sigma x_4^2 \end{bmatrix} = \begin{bmatrix} 20 & 0 & 0 & 0 & 0 \\ 0 & 50 & 0 & 0 & 0 \\ 0 & 0 & 70 & 0 & 0 \\ 0 & 0 & 0 & 50 & 0 \\ 0 & 0 & 0 & 0 & 350 \end{bmatrix}$$

Because $X'X$ is a diagonal matrix, the inverse, $(X'X)^{-1}$, is easily obtained and is simply

$$(X'X)^{-1} = \begin{bmatrix} 1/20 & 0 & 0 & 0 & 0 \\ 0 & 1/50 & 0 & 0 & 0 \\ 0 & 0 & 1/70 & 0 & 0 \\ 0 & 0 & 0 & 1/50 & 0 \\ 0 & 0 & 0 & 0 & 1/350 \end{bmatrix}$$

TABLE 14.3 Values of Y from Table 14.2 with x Variables Consisting of the Coefficients for the Linear (x_1), Quadratic (x_2), Cubic (x_3), and Quartic (x_4) Components of the Trend

Y	X_0	x_1	x_2	x_3	x_4
10	1	-2	2	-1	1
8	1	-2	2	-1	1
12	1	-2	2	-1	1
11	1	-2	2	-1	1
9	1	-2	2	-1	1
12	1	-1	-1	2	-4
14	1	-1	-1	2	-4
11	1	-1	-1	2	-4
10	1	-1	-1	2	-4
13	1	-1	-1	2	-4
9	1	0	-2	0	6
13	1	0	-2	0	6
12	1	0	-2	0	6
10	1	0	-2	0	6
11	1	0	-2	0	6
14	1	1	-1	-2	-4
13	1	1	-1	-2	-4
11	1	1	-1	-2	-4
12	1	1	-1	-2	-4
15	1	1	-1	-2	-4
15	1	2	2	1	1
13	1	2	2	1	1
12	1	2	2	1	1
16	1	2	2	1	1
14	1	2	2	1	1

For $\mathbf{X'Y}$ we have

$$\mathbf{X'Y} = \begin{bmatrix} \Sigma Y \\ \Sigma x_1 Y \\ \Sigma x_2 Y \\ \Sigma x_3 Y \\ \Sigma x_4 Y \end{bmatrix} = \begin{bmatrix} 300 \\ 45 \\ 5 \\ 10 \\ -50 \end{bmatrix}$$

Then for $(\mathbf{X'X})^{-1}\mathbf{X'Y}$, we obtain

$$
(\mathbf{X'X})^{-1}\mathbf{X'Y} =
\begin{bmatrix}
300/20 \\
45/50 \\
5/70 \\
10/50 \\
-50/350
\end{bmatrix}
=
\begin{bmatrix}
12.00 \\
.90 \\
.07 \\
.20 \\
-.14
\end{bmatrix}
= \mathbf{b}
$$

where the values in $\mathbf{b}$ have been rounded. Observe that they are the same as those we obtained previously.

Because the X variables are mutually orthogonal, we have

$$
SS_{reg} = \frac{(\Sigma x_1 Y)^2}{\Sigma x_1^2} + \frac{(\Sigma x_2 Y)^2}{\Sigma x_2^2} + \frac{(\Sigma x_3 Y)^2}{\Sigma x_3^2} + \frac{(\Sigma x_4 Y)^2}{\Sigma x_4^2}
$$

or

$$
SS_{reg} = \frac{(45)^2}{50} + \frac{(5)^2}{70} + \frac{(10)^2}{50} + \frac{(-50)^2}{350} = 50.00
$$

rounded and this is equal to SS_T. Then, we have

$$
R_{YY'}^2 = \frac{SS_{reg}}{SS_{tot}} = \frac{50.00}{100.00} = .50
$$

and .50 of the total sum of squares can be accounted for by the four X variables. Observe that a test of significance of $R_{YY'}^2$ will result in the same value of F as the F test of MS_T.

Note also that if we divide each value of $(\Sigma x_i Y)^2 / \Sigma x_i^2$ by SS_{tot}, we have

$$
R_{YY'}^2 = r_{Y1}^2 + r_{Y2}^2 + r_{Y3}^2 + r_{Y4}^2
$$

$$
= .405 + .004 + .020 + .071
$$

$$
= .50
$$

because the X variables are mutually orthogonal.

With mutually orthogonal X variables, SS_{reg} can be further subdivided into the mutually independent regression sums of squares for the linear, quadratic, cubic, and quartic terms or

$$
SS_{reg} = 40.50 + .36 + 2.00 + 7.14 = 50.00
$$

as we showed earlier in the chapter.

Exercises

14.1 We have $k = 5$ equally spaced values of a variable X with $n = 5$ observations of a dependent variable Y for each value of X. The ordered Y means are 3, 4, 5, 6, and 7. (a) Calculate SS_T and the linear component of SS_T. (b) Calculate the values of $\overline{Y}'_i = \overline{Y} + b_1 x_1$ and show that the correlation between $\overline{Y}'_i$ and $\overline{Y}_i$ is equal to 1.00.

14.2 Assume that the ordered values of the $k = 5$ means in Exercise 14.1 are 7, 4, 3, 4, and 7. (a) Calculate SS_T and the quadratic component of SS_T. (b) Calculate the values of $\overline{Y}'_i = \overline{Y} + b_2 x_2$ and show that the correlation between $\overline{Y}'_i$ and $\overline{Y}_i$ is equal to 1.00.

14.3 Assume that the ordered values of the $k = 5$ means in Exercise 14.1 are 4, 7, 5, 3, and 6. (a) Calculate SS_T and the cubic component of SS_T. (b) Calculate the values of $\overline{Y}'_i = \overline{Y} + b_3 x_3$ and show that the correlation between $\overline{Y}'_i$ and $\overline{Y}_i$ is equal to 1.00.

14.4 What properties do the coefficients for orthogonal polynomials possess? For example, what is true about Σx_i? What is true about $\Sigma x_i x_j, i \neq j$?

14.5 We have $k = 5$ equally spaced values of an independent variable X and $n_i = 5$ observations of a dependent variable for each value of X, as shown in the following table. Make all tests of significance with $\alpha = .05$.

X_1	X_2	X_3	X_4	X_5
3	7	8	4	10
5	9	10	6	9
2	8	6	5	11
1	6	7	8	8
4	5	9	7	7

(a) Calculate SS_T and SS_W and determine whether MS_T is significant. (b) What are the values of b_1, b_2, b_3, and b_4? (c) Calculate the linear, quadratic, cubic, and quartic components of SS_T and test each for significance. (d) Are the deviations from linearity significant? (e) Are the deviations from $\overline{Y}'_i = \overline{Y} + b_1 x_1 + b_2 x_2$ significant? (f) Are the deviations from $\overline{Y}'_i = \overline{Y} + b_1 x_1 + b_2 x_2 + b_3 x_3$ significant? (g) Calculate the squares of the correlation coefficients between $\overline{Y}_i$ and each of the sets of orthogonal coefficients. (h) What proportion of SS_T can be accounted for by the linear and cubic components? (i) Calculate the values of $\overline{Y}'_i = \overline{Y} + b_1 x_1 + b_2 x_2 + b_3 x_3 + b_4 x_4$.

14.6 Given that $\overline{Y}'_i = \overline{Y} + b_1 x_1$, prove that the residual sum of squares

$$SS_{res} = n_i \Sigma (\overline{Y}_i - \overline{Y}'_i)^2$$

is equal to $SS_T - SS_L$.

14.7 Assume that we have a completely randomized design with three equally spaced values of X and that five subjects are assigned at random to each value of X. The Y values for each subject are as follows:

	X_1	X_2	X_3
	3	7	6
	1	4	2
	2	3	4
	4	6	3
	5	5	5
Σ	15	25	20

(a) Find the values of MS_W and MS_T and determine whether the means differ significantly. (b) Find the linear and quadratic components of the treatment sum of squares. (c) Will the quadratic component be equal to the sum of squares for deviations from linearity? Explain why or why not. (d) Is the linear component significant? (e) Are there significant deviations from linearity?

Answers to
Selected Exercises

Answers to exercises involving calculations were, for the most part, obtained with a hand calculator using the full capacity of eight digits. Rounding errors are inevitable when calculations exceed the capacity of the calculator. Final answers have also been rounded. You should keep this in mind in comparing your answers with those reported here. All inverses of matrices 3×3 or larger were obtained using a computer. These inverses are provided in the exercises, and you do not have to calculate any inverse other than those for a 2×2 matrix. Subsequent calculations involving the inverse were done with a hand calculator.

Chapter 1

1.1 $a = 1.8$ and $b = .4$
1.2 $a = -5.0$ and $b = -.4$
1.3 $a = 10$ and $b = .3$
1.4 $a = 5$ and $b = -.2$
1.5 $a = 10$ and $b = -.5$

Chapter 2

2.1 $a = 3.62$ and $b = .483$
2.2 (a) $Y' = -.82 + 1.11X$
        (b) $s_{Y.X}^2 = .855$ and $s_{Y.X} = .925$
        (c) $\overline{Y} = .40$ and $\overline{X} = 1.10$
        (d) $s_Y^2 = 15.60$ and $s_X^2 = 12.10$

Chapter 3

3.1 (a) $\overline{X} = 14.6$ and $\overline{Y} = 9.9$
        (b) $s_X^2 = 81.82$ and $s_Y^2 = 36.54$
        (c) $r = .89$
        (d) $Y' = 1.29 + .59X$
        (e) $X' = 1.53 + 1.32Y$
        (f) $s_{Y.X}^2 = 8.90$ and $s_{X.Y}^2 = 19.94$
3.2 (a) $r = .89$
        (b) $b_Y = 1.00$ and $b_X = .79$
3.4 We have $Y - Y' = y - bx$. Then the numerator of the correlation coefficient will be

$$\Sigma(y - bx)x = \Sigma xy - b\Sigma x^2$$

Substituting $\Sigma xy/\Sigma x^2$ for b in the preceding expression, we see that the numerator of the correlation coefficient is equal to zero.
3.6 $b_X = .75$ and $b_Y = .48$
3.7 $\Sigma(Y - Y')^2 = 0$

3.8   No, because $b_Y b_X = r^2$ and r^2 cannot be greater than 1.00.
3.9   If $r = 1.00$, then $b_X = .50$.
3.10 No, because b_X and b_Y must have the same sign.

Chapter 4

4.5   (a) No, because $r_{12.3} = 2.53$ and this is an impossible value.
      (b) $r_{12} = .02$
      (c) $r_{12} = -1.00$

Chapter 5

5.1 $r = .23$
5.2 $r = .63$
5.3 $r = -.10$
5.4 $r = .13$
5.5 $r = .30$

Chapter 6

6.1 $r \geq .632$ or $r \leq - .632$ would be significant.
6.2 $r \geq .44$ or $r \leq - .44$ would be significant.
6.3 $.63 < \rho < .92$
6.4 $Z = .685$ and $\chi^2 = .470$
6.5   (a) $\chi^2 - 7.611$ with 2 d.f. is a significant value with $\alpha = .05$.
      (b) The estimate would not be obtained because the evidence
      indicates that the samples are not from a common population.

Chapter 7

7.1   (a) .18, .42, .12, and .28
      (b) 9, 21, 6, and 14
      (c) $\chi^2 = 19.44$ with 1 d.f.
7.2   (a) $t = 3.0$ with 14 d.f.
      (b) $F = 9.0$ with 1 and 14 d.f.
7.3   Yes, either from Table IX or from Table V.
7.4 $\chi^2 = (200)(.23)^2 = 10.58$ with 1 d.f.

Chapter 8

8.1   (a) $b = 2.0$
      (b) $s_{Y.X}^2 = 1.3333$ and $s_b = .365$
      (c) $t = 5.48$ with 3 d.f.

8.2 (a) $b = .5$
(b) $s_{Y.X}^2 = .5000$ and $s_b = .224$
(c) $t = 2.236$ with 3 d.f.

8.3 (a) $s_{Y.X}^2 = .9167$
(b) $s_{b_1 - b_2} = .4282$
(c) $t = 3.503$ with 6 d.f.

8.4 (a) $SS_1 = 5.50$, $SS_2 = 16.75$, $SS_3 = 11.25$
(b) $F = 12.27$ with 1 and 6 d.f.
(c) $t^2 = (3.503)^2 = 12.27$

8.8 We have

$$t = \frac{b}{\sqrt{s_{Y.X}^2 / \Sigma x^2}}$$

Multiplying the numerator and the denominator of the right side of the equation by $\sqrt{\Sigma x^2}$ and substituting identities for b and $s_{Y.X}^2$, we obtain

$$t = \frac{\dfrac{\Sigma xy \sqrt{\Sigma x^2}}{\Sigma x^2}}{\sqrt{\left[\Sigma y^2 - \dfrac{(\Sigma xy)^2}{\Sigma x^2}\right] / (n - 2)}}$$

But $\Sigma x^2 = \sqrt{\Sigma x^2} \sqrt{\Sigma x^2}$. Substituting this identity in the numerator of the preceding expression and multiplying both the numerator and the denominator by $1/\sqrt{\Sigma y^2}$, we obtain

$$t = \frac{r}{\sqrt{1 - r^2}} \sqrt{n - 2}$$

Chapter 9

9.1 (a) $b_1 = 2.8$ and $b_2 = 1.2$
(b) $b_1 = 2.8$
(c) $b_2 = 1.2$
(d) $R_{YY'}^2 = .85$
(e) $F = 48.17$ with 2 and 17 d.f.
(f) .49
(g) .36
(h) .85

9.2 (a) $r_{Y1} = .50$ and $r_{Y2} = .50$
(b) Yes
(c) Yes
(d) Yes
(e) $R_{YY'}^2 = .3333$

      (f) $F = 4.25$ with 2 and 17 d.f.

      (g) $F = 2.13$ with 1 and 17 d.f.

      (h) $F = 2.13$ with 1 and 17 d.f.

Chapter 10

10.9   (a) $a = .6$ and $b = .4$

Chapter 11

11.1   (c) $SS_{reg} = 104.635$ and $SS_{res} = 23.765$

        (d) $R^2_{YY'} = .815$ and $F = 8.805$ with 3 and 6 d.f.

        (e) $b_1 = 1.4898$, $b_2 = .1541$, and $b_3 = .2472$

Chapter 12

12.1   See text, Section 12.1.

12.2   See text, Section 12.1.

12.3   With one X variable, we have $b = \Sigma xy/\Sigma x^2$ or with standardized variables $\tilde{b} = \Sigma z_X z_y/\Sigma z_X^2$. But $\Sigma z_X^2 = n - 1$ and, consequently, $\Sigma z_X z_Y/(n - 1) = r_{YX}$.

12.4   With one X variable and with X and Y in standardized form, we have $y' = r_{YX} z_X$, as Exercise 12.3 shows. Then, $\Sigma y'^2/(n - 1) = r_{YX}^2 \Sigma z_X^2/(n - 1) = r_{YX}^2$ because $\Sigma z_X^2/(n - 1) = 1$.

12.6   (b) $\tilde{b}_1 = .6758$, $\tilde{b}_2 = .1177$, and $\tilde{b}_3 = .1895$

Chapter 13

13.1   (a) $SS_T = 90.0$ and $SS_W = 54.0$

        (d) See text, Section 13.3

        (e) See text, Section 13.3.

        (g) $SS_{reg} = SS_T$ and $SS_{res} = SS_W$

        (h) $R^2_{YY'} = SS_{reg}/SS_{tot}$

13.2   (c) See text, Section 13.4.

        (d) See text, Section 13.4.

        (f) $SS_{reg} = SS_T$ and $SS_{res} = SS_W$

        (g) $R^2_{YY'} = SS_{reg}/SS_{tot}$

13.3   (a) $SS_W = 22.00$ and $SS_T = 90.92$

        (e) $SS_{res} = SS_W$ and $SS_{reg} = SS_T$

        (f) $R^2_{YY'} = SS_{reg}/SS_{tot}$

13.4   (d) $SS_{res} = SS_W$ and $SS_{reg} = SS_T$

        (e) $R^2_{YY'} = SS_{reg}/SS_{tot}$

Chapter 14

14.1   (a) $SS_T = 50.0$
$SS_L = 50.0$

14.2   (a) $SS_T = 70.0$
$SS_Q = 70.0$

14.3   (a) $SS_T = 50.0$
$SS_C = 50.0$

14.5   (a) $SS_T = 106.00$, $SS_W = 50.00$, $F = 10.60$ with 4 and 20 d.f.
(b) $b_1 = 1.100$, $b_2 = -.357$, $b_3 = .800$, $b_4 = .114$
(c) Linear $= 60.50$, $F = 24.20$ with 1 and 20 d.f.
Quadratic $= 8.93$, $F = 3.57$ with 1 and 20 d.f.
Cubic $= 32.00$, $F = 12.80$ with 1 and 20 d.f.
Quartic $= 4.57$, $F = 1.83$ with 1 and 20 d.f.
(d) $SS_{res} = 106.00 - 60.50 = 45.50$
$F = 6.07$ with 3 and 20 d.f.
(e) $SS_{res} = 106.00 - 60.50 - 8.93 = 36.57$
$F = 7.31$ with 2 and 20 d.f.
(f) $SS_{res} = 106.00 - 60.50 - 8.93 - 32.00 = 4.57$
$F = 1.83$ with 1 and 20 d.f.
(g) .571, .084, .302, .043
(h) $.571 + .302 = .873$

14.6   If $\overline{Y}_i' = \overline{Y} + b_1 x_1$, then
$$n_i \Sigma(\overline{Y}_i - \overline{Y}_i')^2 = n_i \Sigma[(\overline{Y}_i - \overline{Y}) - b_1 x_1]^2$$
$$= n_i[\Sigma(\overline{Y}_i - \overline{Y})^2 - 2b_1 \Sigma x_1(\overline{Y}_i - \overline{Y}) + b_1^2 \Sigma x_1^2]$$

But

$$\Sigma x_1(\overline{Y}_i - \overline{Y}) = \frac{1}{n_i} \Sigma(x_1 \Sigma Y_i) \text{ and } b_1 = \frac{\Sigma(x_1 \Sigma Y_i)}{n_i \Sigma x_1^2}$$

Then

$$n_i \Sigma(\overline{Y}_i - \overline{Y}_i')^2 = n_i \Sigma(\overline{Y}_i - \overline{Y})^2 - 2\frac{[\Sigma(x_1 \Sigma Y_i)]^2}{n_i \Sigma x_1^2} + \frac{[\Sigma(x_1 \Sigma Y_i)]^2}{n_i \Sigma x_1^2}$$

The first term on the right is the treatment sum of squares, and we also have

$$SS_L = \frac{[\Sigma(x_1 \Sigma Y_i)]^2}{n_i \Sigma x_1^2}$$

so that

$$n_i \Sigma(\overline{Y}_i - \overline{Y}_i')^2 = SS_T - SS_L$$

14.7   (a) $MS_W = 2.5$, $MS_T = 5.0$, and $F = 2.0$
             (b) $SS_L = 2.5$ and $SS_Q = 7.5$
             (c) Yes
             (d) $F = 1.0$ with 1 and 12 d.f.
             (e) $F = 3.0$ with 1 and 12 d.f.

Appendix

TABLE I Areas and Ordinates of the Normal Curve in Terms of $Z = (X - \mu)/\sigma$

(1) Z	(2) A Area from Mean to Z	(3) B Area in Larger Portion	(4) C Area in Smaller Portion	(5) y Ordinate at Z
0.00	.0000	.5000	.5000	.3989
0.01	.0040	.5040	.4960	.3989
0.02	.0080	.5080	.4920	.3989
0.03	.0120	.5120	.4880	.3988
0.04	.0160	.5160	.4840	.3986
0.05	.0199	.5199	.4801	.3984
0.06	.0239	.5239	.4761	.3982
0.07	.0279	.5279	.4721	.3980
0.08	.0319	.5319	.4681	.3977
0.09	.0359	.5359	.4641	.3973
0.10	.0398	.5398	.4602	.3970
0.11	.0438	.5438	.4562	.3965
0.12	.0478	.5478	.4522	.3961
0.13	.0517	.5517	.4483	.3956
0.14	.0557	.5557	.4443	.3951
0.15	.0596	.5596	.4404	.3945
0.16	.0636	.5636	.4364	.3939
0.17	.0675	.5675	.4325	.3932
0.18	.0714	.5714	.4286	.3925
0.19	.0753	.5753	.4247	.3918
0.20	.0793	.5793	.4207	.3910
0.21	.0832	.5832	.4168	.3902
0.22	.0871	.5871	.4129	.3894
0.23	.0910	.5910	.4090	.3885
0.24	.0948	.5948	.4052	.3876
0.25	.0987	.5987	.4013	.3867
0.26	.1026	.6026	.3974	.3857
0.27	.1064	.6064	.3936	.3847
0.28	.1103	.6103	.3897	.3836
0.29	.1141	.6141	.3859	.3825
0.30	.1179	.6179	.3821	.3814
0.31	.1217	.6217	.3783	.3802
0.32	.1255	.6255	.3745	.3790
0.33	.1293	.6293	.3707	.3778
0.34	.1331	.6331	.3669	.3765
0.35	.1368	.6368	.3632	.3752
0.36	.1406	.6406	.3594	.3739
0.37	.1443	.6443	.3557	.3725
0.36	.1480	.6480	.3520	.3712
0.39	.1517	.6517	.3483	.3697
0.40	.1554	.6554	.3446	.3683
0.41	.1591	.6591	.3409	.3668
0.42	.1628	.6628	.3372	.3653
0.43	.1664	.6664	.3336	.3637
0.44	.1700	.6700	.3300	.3621

TABLE I (continued)

(1)	(2)	(3)	(4)	(5)
		B	C	
	A	Area in	Area in	y
	Area from	Larger	Smaller	Ordinate
Z	Mean to Z	Portion	Portion	at Z
0.45	.1736	.6736	.3264	.3605
0.46	.1772	.6772	.3228	.3589
0.47	.1808	.6808	.3192	.3572
0.48	.1844	.6844	.3156	.3555
0.49	.1879	.6879	.3121	.3538
0.50	.1915	.6915	.3085	.3521
0.51	.1950	.6950	.3050	.3503
0.52	.1985	.6985	.3015	.3485
0.53	.2019	.7019	.2981	.3467
0.54	.2054	.7054	.2946	.3448
0.55	.2088	.7088	.2912	.3429
0.56	.2123	.7123	.2877	.3410
0.57	.2157	.7157	.2843	.3391
0.58	.2190	.7190	.2810	.3372
0.59	.2224	.7224	.2776	.3352
0.60	.2257	.7257	.2743	.3332
0.61	.2291	.7291	.2709	.3312
0.62	.2324	.7324	.2676	.3292
0.63	.2357	.7357	.2643	.3271
0.64	.2389	.7389	.2611	.3251
0.65	.2422	.7422	.2578	.3230
0.66	.2454	.7454	.2546	.3209
0.67	.2486	.7486	.2514	.3187
0.68	.2517	.7517	.2483	.3166
0.69	.2549	.7549	.2451	.3144
0.70	.2580	.7580	.2420	.3123
0.71	.2611	.7611	.2389	.3101
0.72	.2642	.7642	.2358	.3079
0.73	.2673	.7673	.2327	.3056
0.74	.2704	.7704	.2296	.3034
0.75	.2734	.7734	.2266	.3011
0.76	.2764	.7764	.2236	.2989
0.77	.2794	.7794	.2206	.2966
0.78	.2823	.7823	.2177	.2943
0.79	.2852	.7852	.2148	.2920
0.80	.2881	.7881	.2119	.2897
0.81	.2910	.7910	.2090	.2874
0.82	.2939	.7939	.2061	.2850
0.83	.2967	.7967	.2033	.2827
0.84	.2995	.7995	.2005	.2803
0.85	.3023	.8023	.1977	.2780
0.86	.3051	.8051	.1949	.2756
0.87	.3078	.8078	.1922	.2732
0.88	.3106	.8106	.1894	.2709
0.89	.3133	.8133	.1867	.2685

TABLE I   (continued)

(1) Z	(2) A Area from Mean to Z	(3) B Area in Larger Portion	(4) C Area in Smaller Portion	(5) y Ordinate at Z
0.90	.3159	.8159	.1841	.2661
0.91	.3186	.8186	.1814	.2637
0.92	.3212	.8212	.1788	.2613
0.93	.3238	.8238	.1762	.2589
0.94	.3264	.8264	.1736	.2565
0.95	.3289	.8289	.1711	.2541
0.96	.3315	.8315	.1685	.2516
0.97	.3340	.8340	.1660	.2492
0.98	.3365	.8365	.1635	.2468
0.99	.3389	.8389	.1611	.2444
1.00	.3413	.8413	.1587	.2420
1.01	.3438	.8438	.1562	.2396
1.02	.3461	.8461	.1539	.2371
1.03	.3485	.8485	.1515	.2347
1.04	.3508	.8508	.1492	.2323
1.05	.3531	.8531	.1469	.2299
1.06	.3554	.8554	.1446	.2275
1.07	.3577	.8577	.1423	.2251
1.08	.3599	.8599	.1401	.2227
1.09	.3621	.8621	.1379	.2203
1.10	.3643	.8643	.1357	.2179
1.11	.3665	.8665	.1335	.2155
1.12	.3686	.8686	.1314	.2131
1.13	.3708	.8708	.1292	.2107
1.14	.3729	.8729	.1271	.2083
1.15	.3749	.8749	.1251	.2059
1.16	.3770	.8770	.1230	.2036
1.17	.3790	.8790	.1210	.2012
1.18	.3810	.8810	.1190	.1989
1.19	.3830	.8830	.1170	.1965
1.20	.3849	.8849	.1151	.1942
1.21	.3869	.8869	.1131	.1919
1.22	.3888	.8888	.1112	.1895
1.23	.3907	.8907	.1093	.1872
1.24	.3925	.8925	.1075	.1849
1.25	.3944	.8944	.1056	.1826
1.26	.3962	.8962	.1038	.1804
1.27	.3980	.8980	.1020	.1781
1.28	.3997	.8997	.1003	.1758
1.29	.4015	.9015	.0985	.1736
1.30	.4032	.9032	.0968	.1714
1.31	.4049	.9049	.0951	.1691
1.32	.4066	.9066	.0934	.1669
1.33	.4082	.9082	.0918	.1647
1.34	.4099	.9099	.0901	.1626

TABLE I (continued)

(1) Z	(2) A Area from Mean to Z	(3) B Area in Larger Portion	(4) C Area in Smaller Portion	(5) y Ordinate at Z
1.35	.4115	.9115	.0885	.1604
1.36	.4131	.9131	.0869	.1582
1.37	.4147	.9147	.0853	.1561
1.38	.4162	.9162	.0838	.1539
1.39	.4177	.9177	.0823	.1518
1.40	.4192	.9192	.0808	.1497
1.41	.4207	.9207	.0793	.1476
1.42	.4222	.9222	.0778	.1456
1.43	.4236	.9236	.0764	.1435
1.44	.4251	.9251	.0749	.1415
1.45	.4265	.9265	.0735	.1394
1.46	.4279	.9279	.0721	.1374
1.47	.4292	.9292	.0708	.1354
1.48	.4306	.9306	.0694	.1334
1.49	.4319	.9319	.0681	.1315
1.50	.4332	.9332	.0668	.1295
1.51	.4345	.9345	.0655	.1276
1.52	.4357	.9357	.0643	.1257
1.53	.4370	.9370	.0630	.1238
1.54	.4382	.9382	.0618	.1219
1.55	.4394	.9394	.0606	.1200
1.56	.4406	.9406	.0594	.1182
1.57	.4418	.9418	.0582	.1163
1.58	.4429	.9429	.0571	.1145
1.59	.4441	.9441	.0559	.1127
1.60	.4452	.9452	.0548	.1109
1.61	.4463	.9463	.0537	.1092
1.62	.4474	.9474	.0526	.1074
1.63	.4484	.9484	.0516	.1057
1.64	.4495	.9495	.0505	.1040
1.65	.4505	.9505	.0495	.1023
1.66	.4515	.9515	.0485	.1006
1.67	.4525	.9525	.0475	.0989
1.68	.4535	.9535	.0465	.0973
1.69	.4545	.9545	.0455	.0957
1.70	.4554	.9554	.0446	.0940
1.71	.4564	.9564	.0436	.0925
1.72	.4573	.9573	.0427	.0909
1.73	.4582	.9582	.0418	.0893
1.74	.4591	.9591	.0409	.0878
1.75	.4599	.9599	.0401	.0863
1.76	.4608	.9608	.0392	.0848
1.77	.4616	.9616	.0384	.0833
1.78	.4625	.9625	.0375	.0818
1.79	.4633	.9633	.0367	.0804

TABLE I (continued)

(1)	(2)	(3) B	(4) C	(5)
Z	A Area from Mean to Z	Area in Larger Portion	Area in Smaller Portion	y Ordinate at Z
1.80	.4641	.9641	.0359	.0790
1.81	.4649	.9649	.0351	.0775
1.82	.4656	.9656	.0344	.0761
1.83	.4664	.9664	.0336	.0748
1.84	.4671	.9671	.0329	.0734
1.85	.4678	.9678	.0322	.0721
1.86	.4686	.9686	.0314	.0707
1.87	.4693	.9693	.0307	.0694
1.88	.4699	.9699	.0301	.0681
1.89	.4706	.9706	.0294	.0669
1.90	.4713	.9713	.0287	.0656
1.91	.4719	.9719	.0281	.0644
1.92	.4726	.9726	.0274	.0632
1.93	.4732	.9732	.0268	.0620
1.94	.4738	.9738	.0262	.0608
1.95	.4744	.9744	.0256	.0596
1.96	.4750	.9750	.0250	.0584
1.97	.4756	.9756	.0244	.0573
1.98	.4761	.9761	.0239	.0562
1.99	.4767	.9767	.0233	.0551
2.00	.4772	.9772	.0228	.0540
2.01	.4778	.9778	.0222	.0529
2.02	.4783	.9783	.0217	.0519
2.03	.4788	.9788	.0212	.0508
2.04	.4793	.9793	.0207	.0498
2.05	.4798	.9798	.0202	.0488
2.06	.4803	.9803	.0197	.0478
2.07	.4808	.9808	.0192	.0468
2.08	.4812	.9812	.0188	.0459
2.09	.4817	.9817	.0183	.0449
2.10	.4821	.9821	.0179	.0440
2.11	.4826	.9826	.0174	.0431
2.12	.4830	.9830	.0170	.0422
2.13	.4834	.9834	.0166	.0413
2.14	.4838	.9838	.0162	.0404
2.15	.4842	.9842	.0158	.0396
2.16	.4846	.9846	.0154	.0387
2.17	.4850	.9850	.0150	.0379
2.18	.4854	.9854	.0146	.0371
2.19	.4857	.9857	.0143	.0363
2.20	.4861	.9861	.0139	.0355
2.21	.4864	.9864	.0136	.0347
2.22	.4868	.9868	.0132	.0339
2.23	.4871	.9871	.0129	.0332
2.24	.4875	.9875	.0125	.0325

TABLE I (continued)

(1)	(2)	(3)	(4)	(5)
		B	C	
	A	Area in	Area in	y
	Area from	Larger	Smaller	Ordinate
Z	Mean to Z	Portion	Portion	at Z
2.25	.4878	.9878	.0122	.0317
2.26	.4881	.9881	.0119	.0310
2.27	.4884	.9884	.0116	.0303
2.28	.4887	.9887	.0113	.0297
2.29	.4890	.9890	.0110	.0290
2.30	.4893	.9893	.0107	.0283
2.31	.4896	.9896	.0104	.0277
2.32	.4898	.9898	.0102	.0270
2.33	.4901	.9901	.0099	.0264
2.34	.4904	.9904	.0096	.0258
2.35	.4906	.9906	.0094	.0252
2.36	.4909	.9909	.0091	.0246
2.37	.4911	.9911	.0089	.0241
2.38	.4913	.9913	.0087	.0235
2.39	.4916	.9916	.0084	.0229
2.40	.4918	.9918	.0082	.0224
2.41	.4920	.9920	.0080	.0219
2.42	.4922	.9922	.0078	.0213
2.43	.4925	.9925	.0075	.0208
2.44	.4927	.9927	.0073	.0203
2.45	.4929	.9929	.0071	.0198
2.46	.4931	.9931	.0069	.0194
2.47	.4932	.9932	.0068	.0189
2.48	.4934	.9934	.0066	.0184
2.49	.4936	.9936	.0064	.0180
2.50	.4938	.9938	.0062	.0175
2.51	.4940	.9940	.0060	.0171
2.52	.4941	.9941	.0059	.0167
2.53	.4943	.9943	.0057	.0163
2.54	.4945	.9945	.0055	.0158
2.55	.4946	.9946	.0054	.0154
2.56	.4948	.9948	.0052	.0151
2.57	.4949	.9949	.0051	.0147
2.58	.4951	.9951	.0049	.0143
2.59	.4952	.9952	.0048	.0139
2.60	.4953	.9953	.0047	.0136
2.61	.4955	.9955	.0045	.0132
2.62	.4956	.9956	.0044	.0129
3.63	.4957	.9957	.0043	.0126
2.64	.4959	.9959	.0041	.0122
2.65	.4960	.9960	.0040	.0119
2.66	.4961	.9961	.0039	.0116
2.67	.4962	.9962	.0038	.0113
2.68	.4963	.9963	.0037	.0110
2.69	.4964	.9964	.0036	.0107

TABLE I (continued)

(1)	(2)	(3) B	(4) C	(5)
Z	A Area from Mean to Z	Area in Larger Portion	Area in Smaller Portion	y Ordinate at Z
2.70	.4965	.9965	.0035	.0104
2.71	.4966	.9966	.0034	.0101
2.72	.4967	.9967	.0033	.0099
2.73	.4968	.9968	.0032	.0096
2.74	.4969	.9969	.0031	.0093
2.75	.4970	.9970	.0030	.0091
2.76	.4971	.9971	.0029	.0088
2.77	.4972	.9972	.0028	.0086
2.78	.4973	.9973	.0027	.0084
2.79	.4974	.9974	.0026	.0081
2.80	.4974	.9974	.0026	.0079
2.81	.4975	.9975	.0025	.0077
2.82	.4976	.9976	.0024	.0075
2.83	.4977	.9977	.0023	.0073
2.84	.4977	.9977	.0023	.0071
2.85	.4978	.9978	.0022	.0069
2.86	.4979	.9979	.0021	.0067
2.87	.4979	.9979	.0021	.0065
2.88	.4980	.9980	.0020	.0063
2.89	.4981	.9981	.0019	.0061
2.90	.4981	.9981	.0019	.0060
2.91	.4982	.9982	.0018	.0058
2.92	.4982	.9982	.0018	.0056
2.93	.4983	.9983	.0017	.0055
2.94	.4984	.9984	.0016	.0053
2.95	.4984	.9984	.0016	.0051
2.96	.4985	.9985	.0015	.0050
2.97	.4985	.9985	.0015	.0048
2.98	.4986	.9986	.0014	.0047
2.99	.4986	.9986	.0014	.0046
3.00	.4987	.9987	.0013	.0044
3.01	.4987	.9987	.0013	.0043
3.02	.4987	.9987	.0013	.0042
3.03	.4988	.9988	.0012	.0040
3.04	.4988	.9988	.0012	.0039
3.05	.4989	.9989	.0011	.0038
3.06	.4989	.9989	.0011	.0037
3.07	.4989	.9989	.0011	.0036
3.08	.4990	.9990	.0010	.0035
3.09	.4990	.9990	.0010	.0034
3.10	.4990	.9990	.0010	.0033
3.11	.4991	.9991	.0009	.0032
3.12	.4991	.9991	.0009	.0031
3.13	.4991	.9991	.0009	.0030
3.14	.4992	.9992	.0008	.0029

TABLE I (continued)

(1)	(2)	(3)	(4)	(5)
		B	C	
	A	Area in	Area in	y
	Area from	Larger	Smaller	Ordinate
Z	Mean to Z	Portion	Portion	at Z
3.15	.4992	.9992	.0008	.0028
3.16	.4992	.9992	.0008	.0027
3.17	.4992	.9992	.0008	.0026
3.18	.4993	.9993	.0007	.0025
3.19	.4993	.9993	.0007	.0025
3.20	.4993	.9993	.0007	.0024
3.21	.4993	.9993	.0007	.0023
3.22	.4994	.9994	.0006	.0022
3.23	.4994	.9994	.0006	.0022
3.24	.4994	.9994	.0006	.0021
3.30	.4995	.9995	.0005	.0017
3.40	.4997	.9997	.0003	.0012
3.50	.4998	.9998	.0002	.0009
3.60	.4998	.9998	.0002	.0006
3.70	.4999	.9999	.0001	.0004

TABLE II Table of χ^2

The probabilities given by the column headings are those for obtaining χ^2 equal to or greater than the tabled value when a null hypothesis is true and when χ^2 has the degrees of freedom given by the column at the left.

Degrees of freedom df	P = .99	.98	.95	.90	.80	.70	.50	.30	.20	.10	.05	.02	.01
1	.000157	.000628	.00393	.0158	.0642	.148	.455	1.074	1.642	2.706	3.841	5.412	6.635
2	.0201	.0404	.103	.211	.446	.713	1.386	2.408	3.219	4.605	5.991	7.824	9.210
3	.115	.185	.352	.584	1.005	1.424	2.366	3.665	4.642	6.251	7.815	9.837	11.341
4	.297	.429	.711	1.064	1.649	2.195	3.357	4.878	5.989	7.779	9.488	11.668	13.277
5	.554	.752	1.145	1.610	2.343	3.000	4.351	6.064	7.289	9.236	11.070	13.388	15.086
6	.872	1.134	1.635	2.204	3.070	3.828	5.348	7.231	8.558	10.645	12.592	15.033	16.812
7	1.239	1.564	2.167	2.833	3.822	4.671	6.346	8.383	9.803	12.017	14.067	16.622	18.475
8	1.646	2.032	2.733	3.490	4.594	5.527	7.344	9.524	11.030	13.362	15.507	18.168	20.090
9	2.088	2.532	3.325	4.168	5.380	6.393	8.343	10.656	12.242	14.684	16.919	19.679	21.666
10	2.558	3.059	3.940	4.865	6.179	7.267	9.342	11.781	13.442	15.987	18.307	21.161	23.209
11	3.053	3.609	4.575	5.578	6.989	8.148	10.341	12.899	14.631	17.275	19.675	22.618	24.725
12	3.571	4.178	5.226	6.304	7.807	9.034	11.340	14.011	15.812	18.549	21.026	24.054	26.217
13	4.107	4.765	5.892	7.042	8.634	9.926	12.340	15.119	16.985	19.812	22.362	25.472	27.688
14	4.660	5.368	6.571	7.790	9.467	10.821	13.339	16.222	18.151	21.064	23.685	26.873	29.141
15	5.229	5.985	7.261	8.547	10.307	11.721	14.339	17.322	19.311	22.307	24.996	28.259	30.578
16	5.812	6.614	7.962	9.312	11.152	12.624	15.338	18.418	20.465	23.542	26.296	29.633	32.000
17	6.408	7.255	8.672	10.085	12.002	13.531	16.338	19.511	21.615	24.769	27.587	30.995	33.409
18	7.015	7.906	9.390	10.865	12.857	14.440	17.338	20.601	22.760	25.989	28.869	32.346	34.805
19	7.633	8.567	10.117	11.651	13.716	15.352	18.338	21.689	23.900	27.204	30.144	33.687	36.191
20	8.260	9.237	10.851	12.443	14.578	16.266	19.337	22.775	25.038	28.412	31.410	35.020	37.566
21	8.897	9.915	11.591	13.240	15.445	17.182	20.337	23.858	26.171	29.615	32.671	36.343	38.932
22	9.542	10.600	12.338	14.041	16.314	18.101	21.337	24.939	27.301	30.813	33.924	37.659	40.289
23	10.196	11.293	13.091	14.848	17.187	19.021	22.337	26.018	28.429	32.007	35.172	38.968	41.638
24	10.856	11.992	13.848	15.659	18.062	19.943	23.337	27.096	29.553	33.196	36.415	40.270	42.980
25	11.524	12.697	14.611	16.473	18.940	20.867	24.337	28.172	30.675	34.382	37.652	41.566	44.314
26	12.198	13.409	15.379	17.292	19.820	21.792	25.336	29.246	31.795	35.563	38.885	42.856	45.642
27	12.879	14.125	16.151	18.114	20.703	22.719	26.336	30.319	32.912	36.741	40.113	44.140	46.963
28	13.565	14.847	16.928	18.939	21.588	23.647	27.336	31.391	34.027	37.916	41.337	45.419	48.278
29	14.256	15.574	17.708	19.768	22.475	24.577	28.336	32.461	35.139	39.087	42.557	46.693	49.588
30	14.953	16.306	18.493	20.599	23.364	25.508	29.336	33.530	36.250	40.256	43.773	47.962	50.892

Source: Reprinted with permission of Macmillan Publishing Company from R. A. Fisher, *Statistical Methods for Research Workers.* Copyright © 1970 University of Adelaide.

TABLE III   Table of *t*

The probabilities given by the column headings are for a one-sided test, assuming a null hypothesis to be true. For example, with 30 d.f., we have $P(t \geq 2.042) = .025$. For a two-sided test, we have $P(t \geq 2.042) + P(t \leq -2.042) = .025 + .025 = .05$.

df \ P	.25	.10	.05	.025	.01	.005	.0025	.001
1	1.000	3.078	6.314	12.706	31.821	63.657	127.321	318.309
2	.816	1.886	2.920	4.303	6.965	9.925	14.089	22.327
3	.765	1.638	2.353	3.182	4.541	5.841	7.453	10.214
4	.741	1.533	2.132	2.776	3.747	4.604	5.598	7.173
5	.727	1.476	2.015	2.571	3.365	4.032	4.773	5.893
6	.718	1.440	1.943	2.447	3.143	3.707	4.317	5.208
7	.711	1.415	1.895	2.365	2.998	3.499	4.029	4.785
8	.706	1.397	1.860	2.306	2.896	3.355	3.833	4.501
9	.703	1.383	1.833	2.262	2.821	3.250	3.690	4.297
10	.700	1.372	1.812	2.228	2.764	3.169	3.581	4.144
11	.697	1.363	1.796	2.201	2.718	3.106	3.497	4.025
12	.695	1.356	1.782	2.179	2.681	3.055	3.428	3.930
13	.694	1.350	1.771	2.160	2.650	3.012	3.372	3.852
14	.692	1.345	1.761	2.145	2.624	2.977	3.326	3.787
15	.691	1.341	1.753	2.131	2.602	2.947	3.286	3.733
16	.690	1.337	1.746	2.120	2.583	2.921	3.252	3.686
17	.689	1.333	1.740	2.110	2.567	2.898	3.223	3.646
18	.688	1.330	1.734	2.101	2.552	2.878	3.197	3.610
19	.688	1.328	1.729	2.093	2.539	2.861	3.174	3.579
20	.687	1.325	1.725	2.086	2.528	2.845	3.153	3.552
21	.686	1.323	1.721	2.080	2.518	2.831	3.135	3.527
22	.686	1.321	1.717	2.074	2.508	2.819	3.119	3.505
23	.685	1.319	1.714	2.069	2.500	2.807	3.104	3.485
24	.685	1.318	1.711	2.064	2.492	2.797	3.090	3.467
25	.684	1.316	1.708	2.060	2.485	2.787	3.078	3.450
26	.684	1.315	1.706	2.056	2.479	2.779	3.067	3.435
27	.684	1.314	1.703	2.052	2.473	2.771	3.057	3.421
28	.683	1.313	1.701	2.048	2.467	2.763	3.047	3.408
29	.683	1.311	1.699	2.045	2.462	2.756	3.038	3.396
30	.683	1.310	1.697	2.042	2.457	2.750	3.030	3.385
35	.682	1.306	1.690	2.030	2.438	2.724	2.996	3.340
40	.681	1.303	1.684	2.021	2.423	2.704	2.971	3.307
45	.680	1.301	1.679	2.014	2.412	2.690	2.952	3.281
50	.679	1.299	1.676	2.009	2.403	2.678	2.937	3.261
55	.679	1.297	1.673	2.004	2.396	2.668	2.925	3.245
60	.679	1.296	1.671	2.000	2.390	2.660	2.915	3.232
70	.678	1.294	1.667	1.994	2.381	2.648	2.899	3.211
80	.678	1.292	1.664	1.990	2.374	2.639	2.887	3.195
90	.677	1.291	1.662	1.987	2.368	2.632	2.878	3.183
100	.677	1.290	1.660	1.984	2.364	2.626	2.871	3.174
200	.676	1.286	1.652	1.972	2.345	2.601	2.838	3.131
500	.675	1.283	1.648	1.965	2.334	2.586	2.820	3.107
1,000	.675	1.282	1.646	1.962	2.330	2.581	2.813	3.098
2,000	.675	1.282	1.645	1.961	2.328	2.578	2.810	3.094
10,000	.675	1.282	1.645	1.960	2.327	2.576	2.808	3.091
∞	.674	1.282	1.645	1.960	2.326	2.576	2.807	3.090

Source: Reprinted from Enrico T. Federighi, Extended tables of the percentage points of student's *t* distribution. *Journal of the American Statistical Association*, 1959, *54*, 683–688, by permission.

TABLE III (continued)

df	.0005	.00025	.0001	.00005	.000025	.00001
1	636.619	1,273.239	3,183.099	6,366.198	12,732.395	31,830.989
2	31.598	44.705	70.700	99.992	141.416	223.603
3	12.924	16.326	22.204	28.000	35.298	47.928
4	8.610	10.306	13.034	15.544	18.522	23.332
5	6.869	7.976	9.678	11.178	12.893	15.547
6	5.959	6.788	8.025	9.082	10.261	12.032
7	5.408	6.082	7.063	7.885	8.782	10.103
8	5.041	5.618	6.442	7.120	7.851	8.907
9	4.781	5.291	6.010	6.594	7.215	8.102
10	4.587	5.049	5.694	6.211	6.757	7.527
11	4.437	4.863	5.453	5.921	6.412	7.098
12	4.318	4.716	5.263	5.694	6.143	6.756
13	4.221	4.597	5.111	5.513	5.928	6.501
14	4.140	4.499	4.985	5.363	5.753	6.287
15	4.073	4.417	4.880	5.239	5.607	6.109
16	4.015	4.346	4.791	5.134	5.484	5.960
17	3.965	4.286	4.714	5.044	5.379	5.832
18	3.922	4.233	4.648	4.966	5.288	5.722
19	3.883	4.187	4.590	4.897	5.209	5.627
20	3.850	4.146	4.539	4.837	5.139	5.543
21	3.819	4.110	4.493	4.784	5.077	5.469
22	3.792	4.077	4.452	4.736	5.022	5.402
23	3.768	4.048	4.415	4.693	4.972	5.343
24	3.745	4.021	4.382	4.654	4.927	5.290
25	3.725	3.997	4.352	4.619	4.887	5.241
26	3.707	3.974	4.324	4.587	4.850	5.197
27	3.690	3.954	4.299	4.558	4.816	5.157
28	3.674	3.935	4.275	4.530	4.784	5.120
29	3.659	3.918	4.254	4.506	4.756	5.086
30	3.646	3.902	4.234	4.482	4.729	5.054
35	3.591	3.836	4.153	4.389	4.622	4.927
40	3.551	3.788	4.094	4.321	4.544	4.835
45	3.520	3.752	4.049	4.269	4.485	4.766
50	3.496	3.723	4.014	4.228	4.438	4.711
55	3.476	3.700	3.986	4.196	4.401	4.667
60	3.460	3.681	3.962	4.169	4.370	4.631
70	3.435	3.651	3.926	4.127	4.323	4.576
80	3.416	3.629	3.899	4.096	4.288	4.535
90	3.402	3.612	3.878	4.072	4.261	4.503
100	3.390	3.598	3.862	4.053	4.240	4.478
200	3.340	3.539	3.789	3.970	4.146	4.369
500	3.310	3.504	3.747	3.922	4.091	4.306
1,000	3.300	3.492	3.733	3.906	4.073	4.285
2,000	3.295	3.486	3.726	3.898	4.064	4.275
10,000	3.292	3.482	3.720	3.892	4.058	4.267
∞	3.291	3.481	3.719	3.891	4.056	4.265

TABLE III   (continued)

df \ P	.000005	.0000025	.000001	.0000005	.00000025	.0000001
1	63,661.977	127,323.954	318,309.886	636,619.772	1,273,239.545	3,183,098.862
2	316.225	447.212	707.106	999.999	1,414.213	2,236.068
3	60.397	76.104	103.299	130.155	163.989	222.572
4	27.771	33.047	41.578	49.459	58.829	73.986
5	17.897	20.591	24.771	28.477	32.734	39.340
6	13.555	15.260	17.830	20.047	22.532	26.286
7	11.215	12.437	14.241	15.764	17.447	19.932
8	9.782	10.731	12.110	13.257	14.504	16.320
9	8.827	9.605	10.720	11.637	12.623	14.041
10	8.150	8.812	9.752	10.516	11.328	12.492
11	7.648	8.227	9.043	9.702	10.397	11.381
12	7.261	7.780	8.504	9.085	9.695	10.551
13	6.955	7.427	8.082	8.604	9.149	9.909
14	6.706	7.142	7.743	8.218	8.713	9.400
15	6.502	6.907	7.465	7.903	8.358	8.986
16	6.330	6.711	7.233	7.642	8.064	8.645
17	6.184	6.545	7.037	7.421	7.817	8.358
18	6.059	6.402	6.869	7.232	7.605	8.115
19	5.949	6.278	6.723	7.069	7.423	7.905
20	5.854	6.170	6.597	6.927	7.265	7.723
21	5.769	6.074	6.485	6.802	7.126	7.564
22	5.694	5.989	6.386	6.692	7.003	7.423
23	5.627	5.913	6.297	6.593	6.893	7.298
24	5.566	5.845	6.218	6.504	6.795	7.185
25	5.511	5.783	6.146	6.424	6.706	7.085
26	5.461	5.726	6.081	6.352	6.626	6.993
27	5.415	5.675	6.021	6.286	6.553	6.910
28	5.373	5.628	5.967	6.225	6.486	6.835
39	5.335	5.585	5.917	6.170	6.426	6.765
30	5.299	5.545	5.871	6.119	6.369	6.701
35	5.156	5.385	5.687	5.915	6.143	6.447
40	5.053	5.269	5.554	5.768	5.983	6.266
45	4.975	5.182	5.454	5.659	5.862	6.130
50	4.914	5.115	5.377	5.573	5.769	6.025
55	4.865	5.060	5.315	5.505	5.694	5.942
60	4.825	5.015	5.264	5.449	5.633	5.873
70	4.763	4.946	5.185	5.363	5.539	5.768
80	4.717	4.896	5.128	5.300	5.470	5.691
90	4.682	4.857	5.084	5.252	5.417	5.633
100	4.654	4.826	5.049	5.214	5.376	5.587
200	4.533	4.692	4.897	5.048	5.196	5.387
500	4.463	4.615	4.810	4.953	5.094	5.273
1,000	4.440	4.590	4.781	4.922	5.060	5.236
2,000	4.428	4.578	4.767	4.907	5.043	5.218
10,000	4.419	4.567	4.756	4.895	5.029	5.203
∞	4.417	4.565	4.753	4.892	5.026	5.199

TABLE IV   Values of the Correlation Coefficient for Various Levels of Significance

The probabilities given by the column headings are for obtaining r equal to or greater than the tabled value, that is, for a one-sided test, when the null hypothesis $\rho = 0$ is true. For a two-sided test, the probabilities should be doubled.

df \ P	.050	.025	.010	.005
1	.988	.997	.9995	.9999
2	.900	.950	.980	.990
3	.805	.878	.934	.959
4	.729	.811	.882	.917
5	.669	.754	.833	.874
6	.622	.707	.789	.834
7	.582	.666	.750	.798
8	.549	.632	.716	.765
9	.521	.602	.685	.735
10	.497	.576	.658	.708
11	.476	.553	.634	.684
12	.458	.532	.612	.661
13	.441	.514	.592	.641
14	.426	.497	.574	.623
15	.412	.482	.558	.606
16	.400	.468	.542	.590
17	.389	.456	.528	.575
18	.378	.444	.516	.561
19	.369	.433	.503	.549
20	.360	.423	.492	.537
21	.352	.413	.482	.526
22	.344	.404	.472	.515
23	.337	.396	.462	.505
24	.330	.388	.453	.496
25	.323	.381	.445	.487
26	.317	.374	.437	.479
27	.311	.367	.430	.471
28	.306	.361	.423	.463
29	.301	.355	.416	.456
30	.296	.349	.409	.449
35	.275	.325	.381	.418
40	.257	.304	.358	.393
45	.243	.288	.338	.372
50	.231	.273	.322	.354
60	.211	.250	.295	.325
70	.195	.232	.274	.302
80	.183	.217	.256	.283
90	.173	.205	.242	.267
100	.164	.195	.230	.254

Additional values of r at the .025 and .005 levels of significance

df	.025	.005	df	.025	.005	df	.025	.005
32	.339	.436	48	.279	.361	150	.159	.208
34	.329	.424	55	.261	.338	175	.148	.193
36	.320	.413	65	.241	.313	200	.138	.181
38	.312	.403	75	.224	.292	300	.113	.148
42	.297	.384	85	.211	.275	400	.098	.128
44	.291	.376	95	.200	.260	500	.088	.115
46	.284	.368	125	.174	.228	1,000	.062	.081

Source: Reprinted with permission of Macmillan Publishing Company from R. A. Fisher, *Statistical Methods for Research Workers.* Copyright © 1970 University of Adelaide. Additional entries were calculated using the t distribution.

TABLE V   Table of Values of $z_r = \frac{1}{2}[\log_e (1 + r) - \log_e(1 - r)]$

r	z_r	r	z_r	r	z_r	r	z_r	r	z_r
.000	.000	.200	.203	.400	.424	.600	.693	.800	1.099
.005	.005	.205	.208	.405	.430	.605	.701	.805	1.113
.010	.010	.210	.213	.410	.436	.610	.709	.810	1.127
.015	.015	.215	.218	.415	.442	.615	.717	.815	1.142
.020	.020	.220	.224	.420	.448	.620	.725	.820	1.157
.025	.025	.225	.229	.425	.454	.625	.733	.825	1.172
.030	.030	.230	.234	.430	.460	.630	.741	.830	1.188
.035	.035	.235	.239	.435	.466	.635	.750	.835	1.204
.040	.040	.240	.245	.440	.472	.640	.758	.840	1.221
.045	.045	.245	.250	.445	.478	.645	.767	.845	1.238
.050	.050	.250	.255	.450	.485	.650	.775	.850	1.256
.055	.055	.255	.261	.455	.491	.655	.784	.855	1.274
.060	.060	.260	.266	.460	.497	.660	.793	.860	1.293
.065	.065	.265	.271	.465	.504	.665	.802	.865	1.313
.070	.070	.270	.277	.470	.510	.670	.811	.870	1.333
.075	.075	.275	.282	.475	.517	.675	.820	.875	1.354
.080	.080	.280	.288	.480	.523	.680	.829	.880	1.376
.085	.085	.285	.293	.485	.530	.685	.838	.885	1.398
.090	.090	.290	.299	.490	.536	.690	.848	.890	1.422
.095	.095	.295	.304	.495	.543	.695	.858	.895	1.447
.100	.100	.300	.310	.500	.549	.700	.867	.900	1.472
.105	.105	.305	.315	.505	.556	.705	.877	.905	1.499
.110	.110	.310	.321	.510	.563	.710	.887	.910	1.528
.115	.116	.315	.326	.515	.570	.715	.897	.915	1.557
.120	.121	.320	.332	.520	.576	.720	.908	.920	1.589
.125	.126	.325	.337	.525	.583	.725	.918	.925	1.623
.130	.131	.330	.343	.530	.590	.730	.929	.930	1.658
.135	.136	.335	.348	.535	.597	.735	.940	.935	1.697
.140	.141	.340	.354	.540	.604	.740	.950	.940	1.738
.145	.146	.345	.360	.545	.611	.745	.962	.945	1.783
.150	.151	.350	.365	.550	.618	.750	.973	.950	1.832
.155	.156	.355	.371	.555	.626	.755	.984	.955	1.886
.160	.161	.360	.377	.560	.633	.760	.996	.960	1.946
.165	.167	.365	.383	.565	.640	.765	1.008	.965	2.014
.170	.172	.370	.388	.570	.648	.770	1.020	.970	2.092
.175	.177	.375	.394	.575	.655	.775	1.033	.975	2.185
.180	.182	.380	.400	.580	.662	.780	1.045	.980	2.298
.185	.187	.385	.406	.585	.670	.785	1.058	.985	2.443
.190	.192	.390	.412	.590	.678	.790	1.071	.990	2.647
.195	.198	.395	.418	.595	.685	.795	1.085	.995	2.994

Source: Table V was constructed by F. P. Kilpatrick and D. A. Buchanan.

TABLE VI   Values of F Significant with $\alpha = .05$ and $\alpha = .01$

The values of F significant at the .05 (roman type) and .01 (boldface type) levels of significance with n_1 degrees of freedom for the numerator and n_2 degrees of freedom for the denominator of the F ratio.

n_1 Degrees of freedom

n_2	1	2	3	4	5	6	7	8	9	10	11	12	14	16	20	24	30	40	50	75	100	200	500	∞
1	161 **4,052**	200 **4,999**	216 **5,403**	225 **5,625**	230 **5,764**	234 **5,859**	237 **5,928**	239 **5,981**	241 **6,022**	242 **6,056**	243 **6,082**	244 **6,106**	245 **6,142**	246 **6,169**	248 **6,208**	249 **6,234**	250 **6,258**	251 **6,286**	252 **6,302**	253 **6,323**	253 **6,334**	254 **6,352**	254 **6,361**	254 **6,366**
2	18.51 **98.49**	19.00 **99.00**	19.16 **99.17**	19.25 **99.25**	19.30 **99.30**	19.33 **99.33**	19.36 **99.34**	19.37 **99.36**	19.38 **99.38**	19.39 **99.40**	19.40 **99.41**	19.41 **99.42**	19.42 **99.43**	19.43 **99.44**	19.44 **99.45**	19.45 **99.46**	19.46 **99.47**	19.47 **99.48**	19.47 **99.48**	19.48 **99.49**	19.49 **99.49**	19.49 **99.49**	19.50 **99.50**	19.50 **99.50**
3	10.13 **34.12**	9.55 **30.82**	9.28 **29.46**	9.12 **28.71**	9.01 **28.24**	8.94 **27.91**	8.88 **27.67**	8.84 **27.49**	8.81 **27.34**	8.78 **27.23**	8.76 **27.13**	8.74 **27.05**	8.71 **26.92**	8.69 **26.83**	8.66 **26.69**	8.64 **26.60**	8.62 **26.50**	8.60 **26.41**	8.58 **26.35**	8.57 **26.27**	8.56 **26.23**	8.54 **26.18**	8.54 **26.14**	8.53 **26.12**
4	7.71 **21.20**	6.94 **18.00**	6.59 **16.69**	6.39 **15.98**	6.26 **15.52**	6.16 **15.21**	6.09 **14.98**	6.04 **14.80**	6.00 **14.66**	5.96 **14.54**	5.93 **14.45**	5.91 **14.37**	5.87 **14.24**	5.84 **14.15**	5.80 **14.02**	5.77 **13.93**	5.74 **13.83**	5.71 **13.74**	5.70 **13.69**	5.68 **13.61**	5.66 **13.57**	5.65 **13.52**	5.64 **13.48**	5.63 **13.46**
5	6.61 **16.26**	5.79 **13.27**	5.41 **12.06**	5.19 **11.39**	5.05 **10.97**	4.95 **10.67**	4.88 **10.45**	4.82 **10.27**	4.78 **10.15**	4.74 **10.05**	4.70 **9.96**	4.68 **9.89**	4.64 **9.77**	4.60 **9.68**	4.56 **9.55**	4.53 **9.47**	4.50 **9.38**	4.46 **9.29**	4.44 **9.24**	4.42 **9.17**	4.40 **9.13**	4.38 **9.07**	4.37 **9.04**	4.36 **9.02**
6	5.99 **13.74**	5.14 **10.92**	4.76 **9.78**	4.53 **9.15**	4.39 **8.75**	4.28 **8.47**	4.21 **8.26**	4.15 **8.10**	4.10 **7.98**	4.06 **7.87**	4.03 **7.79**	4.00 **7.72**	3.96 **7.60**	3.92 **7.52**	3.87 **7.39**	3.84 **7.31**	3.81 **7.23**	3.77 **7.14**	3.75 **7.09**	3.72 **7.02**	3.71 **6.99**	3.69 **6.94**	3.68 **6.90**	3.67 **6.88**
7	5.59 **12.25**	4.74 **9.55**	4.35 **8.45**	4.12 **7.85**	3.97 **7.46**	3.87 **7.19**	3.79 **7.00**	3.73 **6.84**	3.68 **6.71**	3.63 **6.62**	3.60 **6.54**	3.57 **6.47**	3.52 **6.35**	3.49 **6.27**	3.44 **6.15**	3.41 **6.07**	3.38 **5.98**	3.34 **5.90**	3.32 **5.85**	3.29 **5.78**	3.28 **5.75**	3.25 **5.70**	3.24 **5.67**	3.23 **5.65**
8	5.32 **11.26**	4.46 **8.65**	4.07 **7.59**	3.84 **7.01**	3.69 **6.63**	3.58 **6.37**	3.50 **6.19**	3.44 **6.03**	3.39 **5.91**	3.34 **5.82**	3.31 **5.74**	3.28 **5.67**	3.23 **5.56**	3.20 **5.48**	3.15 **5.36**	3.12 **5.28**	3.08 **5.20**	3.05 **5.11**	3.03 **5.06**	3.00 **5.00**	2.98 **4.96**	2.96 **4.91**	2.94 **4.88**	2.93 **4.86**
9	5.12 **10.56**	4.26 **8.02**	3.86 **6.99**	3.63 **6.42**	3.48 **6.06**	3.37 **5.80**	3.29 **5.62**	3.23 **5.47**	3.18 **5.35**	3.13 **5.26**	3.10 **5.18**	3.07 **5.11**	3.02 **5.00**	2.98 **4.92**	2.93 **4.80**	2.90 **4.73**	2.86 **4.64**	2.82 **4.56**	2.80 **4.51**	2.77 **4.45**	2.76 **4.41**	2.73 **4.36**	2.72 **4.33**	2.71 **4.31**
10	4.96 **10.04**	4.10 **7.56**	3.71 **6.55**	3.48 **5.99**	3.33 **5.64**	3.22 **5.39**	3.14 **5.21**	3.07 **5.06**	3.02 **4.95**	2.97 **4.85**	2.94 **4.78**	2.91 **4.71**	2.86 **4.60**	2.82 **4.52**	2.77 **4.41**	2.74 **4.33**	2.70 **4.25**	2.67 **4.17**	2.64 **4.12**	2.61 **4.05**	2.59 **4.01**	2.56 **3.96**	2.55 **3.93**	2.54 **3.91**
11	4.84 **9.65**	3.98 **7.20**	3.59 **6.22**	3.36 **5.67**	3.20 **5.32**	3.09 **5.07**	3.01 **4.88**	2.95 **4.74**	2.90 **4.63**	2.86 **4.54**	2.82 **4.46**	2.79 **4.40**	2.74 **4.29**	2.70 **4.21**	2.65 **4.10**	2.61 **4.02**	2.57 **3.94**	2.53 **3.86**	2.50 **3.80**	2.47 **3.74**	2.45 **3.70**	2.42 **3.66**	2.41 **3.62**	2.40 **3.60**
12	4.75 **9.33**	3.88 **6.93**	3.49 **5.95**	3.26 **5.41**	3.11 **5.06**	3.00 **4.82**	2.92 **4.65**	2.85 **4.50**	2.80 **4.39**	2.76 **4.30**	2.72 **4.22**	2.69 **4.16**	2.64 **4.05**	2.60 **3.98**	2.54 **3.86**	2.50 **3.78**	2.46 **3.70**	2.42 **3.61**	2.40 **3.56**	2.36 **3.49**	2.35 **3.46**	2.32 **3.41**	2.31 **3.38**	2.30 **3.36**
13	4.67 **9.07**	3.80 **6.70**	3.41 **5.74**	3.18 **5.20**	3.02 **4.86**	2.92 **4.62**	2.84 **4.44**	2.77 **4.30**	2.72 **4.19**	2.67 **4.10**	2.63 **4.02**	2.60 **3.96**	2.55 **3.85**	2.51 **3.78**	2.46 **3.67**	2.42 **3.59**	2.38 **3.51**	2.34 **3.42**	2.32 **3.37**	2.28 **3.30**	2.26 **3.27**	2.24 **3.21**	2.22 **3.18**	2.21 **3.16**
14	4.60 **8.86**	3.74 **6.51**	3.34 **5.56**	3.11 **5.03**	2.96 **4.69**	2.85 **4.46**	2.77 **4.28**	2.70 **4.14**	2.65 **4.03**	2.60 **3.94**	2.56 **3.86**	2.53 **3.80**	2.48 **3.70**	2.44 **3.62**	2.39 **3.51**	2.35 **3.43**	2.31 **3.34**	2.27 **3.26**	2.24 **3.21**	2.21 **3.14**	2.19 **3.11**	2.16 **3.06**	2.14 **3.02**	2.13 **3.00**
15	4.54 **8.68**	3.68 **6.36**	3.29 **5.42**	3.06 **4.89**	2.90 **4.56**	2.79 **4.32**	2.70 **4.14**	2.64 **4.00**	2.59 **3.89**	2.55 **3.80**	2.51 **3.73**	2.48 **3.67**	2.43 **3.56**	2.39 **3.48**	2.33 **3.36**	2.29 **3.29**	2.25 **3.20**	2.21 **3.12**	2.18 **3.07**	2.15 **3.00**	2.12 **2.97**	2.10 **2.92**	2.08 **2.89**	2.07 **2.87**
16	4.49 **8.53**	3.63 **6.23**	3.24 **5.29**	3.01 **4.77**	2.85 **4.44**	2.74 **4.20**	2.66 **4.03**	2.59 **3.89**	2.54 **3.78**	2.49 **3.69**	2.45 **3.61**	2.42 **3.55**	2.37 **3.45**	2.33 **3.37**	2.28 **3.25**	2.24 **3.18**	2.20 **3.10**	2.16 **3.01**	2.13 **2.96**	2.09 **2.89**	2.07 **2.86**	2.04 **2.80**	2.02 **2.77**	2.01 **2.75**

Source: Reprinted by permission from *Statistical Methods* (6th ed.), by George W. Snedecor and William G. Cochran. Copyright © 1967 by Iowa State University Press, Ames, Iowa.

TABLE VI (continued)

n_1, Degrees of freedom

n_2	1	2	3	4	5	6	7	8	9	10	11	12	14	16	20	24	30	40	50	75	100	200	500	∞
17	4.45 / 8.40	3.59 / 6.11	3.20 / 5.18	2.96 / 4.67	2.81 / 4.34	2.70 / 4.10	2.62 / 3.93	2.55 / 3.79	2.50 / 3.68	2.45 / 3.59	2.41 / 3.52	2.38 / 3.45	2.33 / 3.35	2.29 / 3.27	2.23 / 3.16	2.19 / 3.08	2.15 / 3.00	2.11 / 2.92	2.08 / 2.86	2.04 / 2.79	2.02 / 2.76	1.99 / 2.70	1.97 / 2.67	1.96 / 2.65
18	4.41 / 8.28	3.55 / 6.01	3.16 / 5.09	2.93 / 4.58	2.77 / 4.25	2.66 / 4.01	2.58 / 3.85	2.51 / 3.71	2.46 / 3.60	2.41 / 3.51	2.37 / 3.44	2.34 / 3.37	2.29 / 3.27	2.25 / 3.19	2.19 / 3.07	2.15 / 3.00	2.11 / 2.91	2.07 / 2.83	2.04 / 2.78	2.00 / 2.71	1.98 / 2.68	1.95 / 2.62	1.93 / 2.59	1.92 / 2.57
19	4.38 / 8.18	3.52 / 5.93	3.13 / 5.01	2.90 / 4.50	2.74 / 4.17	2.63 / 3.94	2.55 / 3.77	2.48 / 3.63	2.43 / 3.52	2.38 / 3.43	2.34 / 3.36	2.31 / 3.30	2.26 / 3.19	2.21 / 3.12	2.15 / 3.00	2.11 / 2.92	2.07 / 2.84	2.02 / 2.76	2.00 / 2.70	1.96 / 2.63	1.94 / 2.60	1.91 / 2.54	1.90 / 2.51	1.88 / 2.49
20	4.35 / 8.10	3.49 / 5.85	3.10 / 4.94	2.87 / 4.43	2.71 / 4.10	2.60 / 3.87	2.52 / 3.71	2.45 / 3.56	2.40 / 3.45	2.35 / 3.37	2.31 / 3.30	2.28 / 3.23	2.23 / 3.13	2.18 / 3.05	2.12 / 2.94	2.08 / 2.86	2.04 / 2.77	1.99 / 2.69	1.96 / 2.63	1.92 / 2.56	1.90 / 2.53	1.87 / 2.47	1.85 / 2.44	1.84 / 2.42
21	4.32 / 8.02	3.47 / 5.78	3.07 / 4.87	2.84 / 4.37	2.68 / 4.04	2.57 / 3.81	2.49 / 3.65	2.42 / 3.51	2.37 / 3.40	2.32 / 3.31	2.28 / 3.24	2.25 / 3.17	2.20 / 3.07	2.15 / 2.99	2.09 / 2.88	2.05 / 2.80	2.00 / 2.72	1.96 / 2.63	1.93 / 2.58	1.89 / 2.51	1.87 / 2.47	1.84 / 2.42	1.82 / 2.38	1.81 / 2.36
22	4.30 / 7.94	3.44 / 5.72	3.05 / 4.82	2.82 / 4.31	2.66 / 3.99	2.55 / 3.76	2.47 / 3.59	2.40 / 3.45	2.35 / 3.35	2.30 / 3.26	2.26 / 3.18	2.23 / 3.12	2.18 / 3.02	2.13 / 2.94	2.07 / 2.83	2.03 / 2.75	1.98 / 2.67	1.93 / 2.58	1.91 / 2.53	1.87 / 2.46	1.84 / 2.42	1.81 / 2.37	1.80 / 2.33	1.78 / 2.31
23	4.28 / 7.88	3.42 / 5.66	3.03 / 4.76	2.80 / 4.26	2.64 / 3.94	2.53 / 3.71	2.45 / 3.54	2.38 / 3.41	2.32 / 3.30	2.28 / 3.21	2.24 / 3.14	2.20 / 3.07	2.14 / 2.97	2.10 / 2.89	2.04 / 2.78	2.00 / 2.70	1.96 / 2.62	1.91 / 2.53	1.88 / 2.48	1.84 / 2.41	1.82 / 2.37	1.79 / 2.32	1.77 / 2.28	1.76 / 2.26
24	4.26 / 7.82	3.40 / 5.61	3.01 / 4.72	2.78 / 4.22	2.62 / 3.90	2.51 / 3.67	2.43 / 3.50	2.36 / 3.36	2.30 / 3.25	2.26 / 3.17	2.22 / 3.09	2.18 / 3.03	2.13 / 2.93	2.09 / 2.85	2.02 / 2.74	1.98 / 2.66	1.94 / 2.58	1.89 / 2.49	1.86 / 2.44	1.82 / 2.36	1.80 / 2.33	1.76 / 2.27	1.74 / 2.23	1.73 / 2.21
25	4.24 / 7.77	3.38 / 5.57	2.99 / 4.68	2.76 / 4.18	2.60 / 3.86	2.49 / 3.63	2.41 / 3.46	2.34 / 3.32	2.28 / 3.21	2.24 / 3.13	2.20 / 3.05	2.16 / 2.99	2.11 / 2.89	2.06 / 2.81	2.00 / 2.70	1.96 / 2.62	1.92 / 2.54	1.87 / 2.45	1.84 / 2.40	1.80 / 2.32	1.77 / 2.29	1.74 / 2.23	1.72 / 2.19	1.71 / 2.17
26	4.22 / 7.72	3.37 / 5.53	2.98 / 4.64	2.74 / 4.14	2.59 / 3.82	2.47 / 3.59	2.39 / 3.42	2.32 / 3.29	2.27 / 3.17	2.22 / 3.09	2.18 / 3.02	2.15 / 2.96	2.10 / 2.86	2.05 / 2.77	1.99 / 2.66	1.95 / 2.58	1.90 / 2.50	1.85 / 2.41	1.82 / 2.36	1.78 / 2.28	1.76 / 2.25	1.72 / 2.19	1.70 / 2.15	1.69 / 2.13
27	4.21 / 7.68	3.35 / 5.49	2.96 / 4.60	2.73 / 4.11	2.57 / 3.79	2.46 / 3.56	2.37 / 3.39	2.30 / 3.26	2.25 / 3.14	2.20 / 3.06	2.16 / 2.98	2.13 / 2.93	2.08 / 2.83	2.03 / 2.74	1.97 / 2.63	1.93 / 2.55	1.88 / 2.47	1.84 / 2.38	1.80 / 2.33	1.76 / 2.25	1.74 / 2.21	1.71 / 2.16	1.68 / 2.12	1.67 / 2.10
28	4.20 / 7.64	3.34 / 5.45	2.95 / 4.57	2.71 / 4.07	2.56 / 3.76	2.44 / 3.53	2.36 / 3.36	2.29 / 3.23	2.24 / 3.11	2.19 / 3.03	2.15 / 2.95	2.12 / 2.90	2.06 / 2.80	2.02 / 2.71	1.96 / 2.60	1.91 / 2.52	1.87 / 2.44	1.81 / 2.35	1.78 / 2.30	1.75 / 2.22	1.72 / 2.18	1.69 / 2.13	1.67 / 2.09	1.65 / 2.06
29	4.18 / 7.60	3.33 / 5.42	2.93 / 4.54	2.70 / 4.04	2.54 / 3.73	2.43 / 3.50	2.35 / 3.33	2.28 / 3.20	2.22 / 3.08	2.18 / 3.00	2.14 / 2.92	2.10 / 2.87	2.05 / 2.77	2.00 / 2.68	1.94 / 2.57	1.90 / 2.49	1.85 / 2.41	1.80 / 2.32	1.77 / 2.27	1.73 / 2.19	1.71 / 2.15	1.68 / 2.10	1.65 / 2.06	1.64 / 2.03
30	4.17 / 7.56	3.32 / 5.39	2.92 / 4.51	2.69 / 4.02	2.53 / 3.70	2.42 / 3.47	2.34 / 3.30	2.27 / 3.17	2.21 / 3.06	2.16 / 2.98	2.12 / 2.90	2.09 / 2.84	2.04 / 2.74	1.99 / 2.66	1.93 / 2.55	1.89 / 2.47	1.84 / 2.38	1.79 / 2.29	1.76 / 2.24	1.72 / 2.16	1.69 / 2.13	1.66 / 2.07	1.64 / 2.03	1.62 / 2.01
32	4.15 / 7.50	3.30 / 5.34	2.90 / 4.46	2.67 / 3.97	2.51 / 3.66	2.40 / 3.42	2.32 / 3.25	2.25 / 3.12	2.19 / 3.01	2.14 / 2.94	2.10 / 2.86	2.07 / 2.80	2.02 / 2.70	1.97 / 2.62	1.91 / 2.51	1.86 / 2.42	1.82 / 2.34	1.76 / 2.25	1.74 / 2.20	1.69 / 2.12	1.67 / 2.08	1.64 / 2.02	1.61 / 1.98	1.59 / 1.96
34	4.13 / 7.44	3.28 / 5.29	2.88 / 4.42	2.65 / 3.93	2.49 / 3.61	2.38 / 3.38	2.30 / 3.21	2.23 / 3.08	2.17 / 2.97	2.12 / 2.89	2.08 / 2.82	2.05 / 2.76	2.00 / 2.66	1.95 / 2.58	1.89 / 2.47	1.84 / 2.38	1.80 / 2.30	1.74 / 2.21	1.71 / 2.15	1.67 / 2.08	1.64 / 2.04	1.61 / 1.98	1.59 / 1.94	1.57 / 1.91
36	4.11 / 7.39	3.26 / 5.25	2.86 / 4.38	2.63 / 3.89	2.48 / 3.58	2.36 / 3.35	2.28 / 3.18	2.21 / 3.04	2.15 / 2.94	2.10 / 2.86	2.06 / 2.78	2.03 / 2.72	1.98 / 2.62	1.93 / 2.54	1.87 / 2.43	1.82 / 2.35	1.78 / 2.26	1.72 / 2.17	1.69 / 2.12	1.65 / 2.04	1.62 / 2.00	1.59 / 1.94	1.56 / 1.90	1.55 / 1.87
38	4.10 / 7.35	3.25 / 5.21	2.85 / 4.34	2.62 / 3.86	2.46 / 3.54	2.35 / 3.32	2.26 / 3.15	2.19 / 3.02	2.14 / 2.91	2.09 / 2.82	2.05 / 2.75	2.02 / 2.69	1.96 / 2.59	1.92 / 2.51	1.85 / 2.40	1.80 / 2.32	1.76 / 2.22	1.71 / 2.14	1.67 / 2.08	1.63 / 2.00	1.60 / 1.97	1.57 / 1.90	1.54 / 1.86	1.53 / 1.84

TABLE VI (continued)

n₁, Degrees of freedom

n_2	1	2	3	4	5	6	7	8	9	10	11	12	14	16	20	24	30	40	50	75	100	200	500	∞
40	4.08 / 7.31	3.23 / 5.18	2.84 / 4.31	2.61 / 3.83	2.45 / 3.51	2.34 / 3.29	2.25 / 3.12	2.18 / 2.99	2.12 / 2.88	2.07 / 2.80	2.04 / 2.73	2.00 / 2.66	1.95 / 2.56	1.90 / 2.49	1.84 / 2.37	1.79 / 2.29	1.74 / 2.20	1.69 / 2.11	1.66 / 2.05	1.61 / 1.97	1.59 / 1.94	1.55 / 1.88	1.53 / 1.84	1.51 / 1.81
42	4.07 / 7.27	3.22 / 5.15	2.83 / 4.29	2.59 / 3.80	2.44 / 3.49	2.32 / 3.26	2.24 / 3.10	2.17 / 2.96	2.11 / 2.86	2.06 / 2.77	2.02 / 2.70	1.99 / 2.64	1.94 / 2.54	1.89 / 2.46	1.82 / 2.35	1.78 / 2.26	1.73 / 2.17	1.68 / 2.08	1.64 / 2.02	1.60 / 1.94	1.57 / 1.91	1.54 / 1.85	1.51 / 1.80	1.49 / 1.78
44	4.06 / 7.24	3.21 / 5.12	2.82 / 4.26	2.58 / 3.78	2.43 / 3.46	2.31 / 3.24	2.23 / 3.07	2.16 / 2.94	2.10 / 2.84	2.05 / 2.75	2.01 / 2.68	1.98 / 2.62	1.92 / 2.52	1.88 / 2.44	1.81 / 2.32	1.76 / 2.24	1.72 / 2.15	1.66 / 2.06	1.63 / 2.00	1.58 / 1.92	1.56 / 1.88	1.52 / 1.82	1.50 / 1.78	1.48 / 1.75
46	4.05 / 7.21	3.20 / 5.10	2.81 / 4.24	2.57 / 3.76	2.42 / 3.44	2.30 / 3.22	2.22 / 3.05	2.14 / 2.92	2.09 / 2.82	2.04 / 2.73	2.00 / 2.66	1.97 / 2.60	1.91 / 2.50	1.87 / 2.42	1.80 / 2.30	1.75 / 2.22	1.71 / 2.13	1.65 / 2.04	1.62 / 1.98	1.57 / 1.90	1.54 / 1.86	1.51 / 1.80	1.48 / 1.76	1.46 / 1.72
48	4.04 / 7.19	3.19 / 5.08	2.80 / 4.22	2.56 / 3.74	2.41 / 3.42	2.30 / 3.20	2.21 / 3.04	2.14 / 2.90	2.08 / 2.80	2.03 / 2.71	1.99 / 2.64	1.96 / 2.58	1.90 / 2.48	1.86 / 2.40	1.79 / 2.28	1.74 / 2.20	1.70 / 2.11	1.64 / 2.02	1.61 / 1.96	1.56 / 1.88	1.53 / 1.84	1.50 / 1.78	1.47 / 1.73	1.45 / 1.70
50	4.03 / 7.17	3.18 / 5.06	2.79 / 4.20	2.56 / 3.72	2.40 / 3.41	2.29 / 3.18	2.20 / 3.02	2.13 / 2.88	2.07 / 2.78	2.02 / 2.70	1.98 / 2.62	1.95 / 2.56	1.90 / 2.46	1.85 / 2.39	1.78 / 2.26	1.74 / 2.18	1.69 / 2.10	1.63 / 2.00	1.60 / 1.94	1.55 / 1.86	1.52 / 1.82	1.48 / 1.76	1.46 / 1.71	1.44 / 1.68
55	4.02 / 7.12	3.17 / 5.01	2.78 / 4.16	2.54 / 3.68	2.38 / 3.37	2.27 / 3.15	2.18 / 2.98	2.11 / 2.85	2.05 / 2.75	2.00 / 2.66	1.97 / 2.59	1.93 / 2.53	1.88 / 2.43	1.83 / 2.35	1.76 / 2.23	1.72 / 2.15	1.67 / 2.06	1.61 / 1.96	1.58 / 1.90	1.52 / 1.82	1.50 / 1.78	1.46 / 1.71	1.43 / 1.66	1.41 / 1.64
60	4.00 / 7.08	3.15 / 4.98	2.76 / 4.13	2.52 / 3.65	2.37 / 3.34	2.25 / 3.12	2.17 / 2.95	2.10 / 2.82	2.04 / 2.72	1.99 / 2.63	1.95 / 2.56	1.92 / 2.50	1.86 / 2.40	1.81 / 2.32	1.75 / 2.20	1.70 / 2.12	1.65 / 2.03	1.59 / 1.93	1.56 / 1.87	1.50 / 1.79	1.48 / 1.74	1.44 / 1.68	1.41 / 1.63	1.39 / 1.60
65	3.99 / 7.04	3.14 / 4.95	2.75 / 4.10	2.51 / 3.62	2.36 / 3.31	2.24 / 3.09	2.15 / 2.93	2.08 / 2.79	2.02 / 2.70	1.98 / 2.61	1.94 / 2.54	1.90 / 2.47	1.85 / 2.37	1.80 / 2.30	1.73 / 2.18	1.68 / 2.09	1.63 / 2.00	1.57 / 1.90	1.54 / 1.84	1.49 / 1.76	1.46 / 1.71	1.42 / 1.64	1.39 / 1.60	1.37 / 1.56
70	3.98 / 7.01	3.13 / 4.92	2.74 / 4.08	2.50 / 3.60	2.35 / 3.29	2.23 / 3.07	2.14 / 2.91	2.07 / 2.77	2.01 / 2.67	1.97 / 2.59	1.93 / 2.51	1.89 / 2.45	1.84 / 2.35	1.79 / 2.28	1.72 / 2.15	1.67 / 2.07	1.62 / 1.98	1.56 / 1.88	1.53 / 1.82	1.47 / 1.74	1.45 / 1.69	1.40 / 1.62	1.37 / 1.56	1.35 / 1.53
80	3.96 / 6.96	3.11 / 4.88	2.72 / 4.04	2.48 / 3.56	2.33 / 3.25	2.21 / 3.04	2.12 / 2.87	2.05 / 2.74	1.99 / 2.64	1.95 / 2.55	1.91 / 2.48	1.88 / 2.41	1.82 / 2.32	1.77 / 2.24	1.70 / 2.11	1.65 / 2.03	1.60 / 1.94	1.54 / 1.84	1.51 / 1.78	1.45 / 1.70	1.42 / 1.65	1.38 / 1.57	1.35 / 1.52	1.32 / 1.49
100	3.94 / 6.90	3.09 / 4.82	2.70 / 3.98	2.46 / 3.51	2.30 / 3.20	2.19 / 2.99	2.10 / 2.82	2.03 / 2.69	1.97 / 2.59	1.92 / 2.51	1.88 / 2.43	1.85 / 2.36	1.79 / 2.26	1.75 / 2.19	1.68 / 2.06	1.63 / 1.98	1.57 / 1.89	1.51 / 1.79	1.48 / 1.73	1.42 / 1.64	1.39 / 1.59	1.34 / 1.51	1.30 / 1.46	1.28 / 1.43
125	3.92 / 6.84	3.07 / 4.78	2.68 / 3.94	2.44 / 3.47	2.29 / 3.17	2.17 / 2.95	2.08 / 2.79	2.01 / 2.65	1.95 / 2.56	1.90 / 2.47	1.86 / 2.40	1.83 / 2.33	1.77 / 2.23	1.72 / 2.15	1.65 / 2.03	1.60 / 1.94	1.55 / 1.85	1.49 / 1.75	1.45 / 1.68	1.39 / 1.59	1.36 / 1.54	1.31 / 1.46	1.27 / 1.40	1.25 / 1.37
150	3.91 / 6.81	3.06 / 4.75	2.67 / 3.91	2.43 / 3.44	2.27 / 3.14	2.16 / 2.92	2.07 / 2.76	2.00 / 2.62	1.94 / 2.53	1.89 / 2.44	1.85 / 2.37	1.82 / 2.30	1.76 / 2.20	1.71 / 2.12	1.64 / 2.00	1.59 / 1.91	1.54 / 1.83	1.47 / 1.72	1.44 / 1.66	1.37 / 1.56	1.34 / 1.51	1.29 / 1.43	1.25 / 1.37	1.22 / 1.33
200	3.89 / 6.76	3.04 / 4.71	2.65 / 3.88	2.41 / 3.41	2.26 / 3.11	2.14 / 2.90	2.05 / 2.73	1.98 / 2.60	1.92 / 2.50	1.87 / 2.41	1.83 / 2.34	1.80 / 2.28	1.74 / 2.17	1.69 / 2.09	1.62 / 1.97	1.57 / 1.88	1.52 / 1.79	1.45 / 1.69	1.42 / 1.62	1.35 / 1.53	1.32 / 1.48	1.26 / 1.39	1.22 / 1.33	1.19 / 1.28
400	3.86 / 6.70	3.02 / 4.66	2.62 / 3.83	2.39 / 3.36	2.23 / 3.06	2.12 / 2.85	2.03 / 2.69	1.96 / 2.55	1.90 / 2.46	1.85 / 2.37	1.81 / 2.29	1.78 / 2.23	1.72 / 2.12	1.67 / 2.04	1.60 / 1.92	1.54 / 1.84	1.49 / 1.74	1.42 / 1.64	1.38 / 1.57	1.32 / 1.47	1.28 / 1.42	1.22 / 1.32	1.16 / 1.24	1.13 / 1.19
1000	3.85 / 6.66	3.00 / 4.62	2.61 / 3.80	2.38 / 3.34	2.22 / 3.04	2.10 / 2.82	2.02 / 2.66	1.95 / 2.53	1.89 / 2.43	1.84 / 2.34	1.80 / 2.26	1.76 / 2.20	1.70 / 2.09	1.65 / 2.01	1.58 / 1.89	1.53 / 1.81	1.47 / 1.71	1.41 / 1.61	1.36 / 1.54	1.30 / 1.44	1.26 / 1.38	1.19 / 1.28	1.13 / 1.19	1.08 / 1.11
∞	3.84 / 6.64	2.99 / 4.60	2.60 / 3.78	2.37 / 3.32	2.21 / 3.02	2.09 / 2.80	2.01 / 2.64	1.94 / 2.51	1.88 / 2.41	1.83 / 2.32	1.79 / 2.24	1.75 / 2.18	1.69 / 2.07	1.64 / 1.99	1.57 / 1.87	1.52 / 1.79	1.46 / 1.69	1.40 / 1.59	1.35 / 1.52	1.28 / 1.41	1.24 / 1.36	1.17 / 1.25	1.11 / 1.15	1.00 / 1.00

TABLE VII   Values of the Rank Order Correlation Coefficient for Various Levels of Significance

The probabilities given are for a one-sided test. For a two-sided test, the probabilities should be doubled.

n	r	P	n	r	P
4	1.000	.0417	8	.810	.0108
			8	.738	.0224
			8	.690	.0331
5	1.000	.0083	8	.643	.0469
5	.900	.0417	8	.619	.0550
5	.800	.0667	8	.595	.0639
5	.700	.1167	8	.524	.0956
			9	.767	.0106
6	.943	.0083	9	.700	.0210
6	.886	.0167	9	.650	.0323
6	.829	.0292	9	.617	.0417
6	.771	.0514	9	.583	.0528
6	.657	.0875	9	.550	.0656
			9	.467	.1058
7	.857	.0119	10	.733	.0100
7	.786	.0240	10	.661	.0210
7	.750	.0331	10	.612	.0324
7	.714	.0440	10	.576	.0432
7	.679	.0548	10	.552	.0515
7	.643	.0694	10	.527	.0609
7	.571	.1000	10	.442	.1021

Source: The values of *r* were calculated from Table IV of E. G. Olds, Distributions of sums of squares of rank differences for small numbers of individuals, *Annals of Mathematical Statistics*, 1939, *9*, 133–148.

TABLE VIII Table of Coefficients for Orthogonal Polynomials

k	Polynomial	Values of the Coefficients									
3	Linear	−1	0	1							
	Quadratic	1	−2	1							
4	Linear	−3	−1	1	3						
	Quadratic	1	−1	−1	1						
	Cubic	−1	3	−3	1						
5	Linear	−2	−1	0	1	2					
	Quadratic	2	−1	−2	−1	2					
	Cubic	−1	2	0	−2	1					
	Quartic	1	−4	6	−4	1					
6	Linear	−5	−3	−1	1	3	5				
	Quadratic	5	−1	−4	−4	−1	5				
	Cubic	−5	7	4	−4	−7	5				
	Quartic	1	−3	2	2	−3	1				
7	Linear	−3	−2	−1	0	1	2	3			
	Quadratic	5	0	−3	−4	−3	0	5			
	Cubic	−1	1	1	0	−1	−1	1			
	Quartic	3	−7	1	6	1	−7	3			
8	Linear	−7	−5	−3	−1	1	3	5	7		
	Quadratic	7	1	−3	−5	−5	−3	1	7		
	Cubic	−7	5	7	3	−3	−7	−5	7		
	Quartic	7	−13	−3	9	9	−3	−13	7		
9	Linear	−4	−3	−2	−1	0	1	2	3	4	
	Quadratic	28	7	−8	−17	−20	−17	−8	7	28	
	Cubic	−14	7	13	9	0	−9	−13	−7	14	
	Quartic	14	−21	−11	9	18	9	−11	−21	14	
10	Linear	−9	−7	−5	−3	−1	1	3	5	7	9
	Quadratic	6	2	−1	−3	−4	−4	−3	−1	2	6
	Cubic	−42	14	35	31	12	−12	−31	−35	−14	42
	Quartic	18	−22	−17	3	18	18	3	−17	−22	18

Index